T0202012

Listening to British Nature

Listening to British Nature

Wartime, Radio, and Modern Life, 1914–1945

MICHAEL GUIDA

OXFORD
UNIVERSITY PRESS

OXFORD
UNIVERSITY PRESS

Oxford University Press is a department of the University of Oxford. It furthers the University's objective of excellence in research, scholarship, and education by publishing worldwide. Oxford is a registered trade mark of Oxford University Press in the UK and certain other countries.

Published in the United States of America by Oxford University Press
198 Madison Avenue, New York, NY 10016, United States of America.

© Oxford University Press 2022

All rights reserved. No part of this publication may be reproduced, stored in a retrieval system, or transmitted, in any form or by any means, without the prior permission in writing of Oxford University Press, or as expressly permitted by law, by license, or under terms agreed with the appropriate reproduction rights organization. Inquiries concerning reproduction outside the scope of the above should be sent to the Rights Department, Oxford University Press, at the address above.

You must not circulate this work in any other form
and you must impose this same condition on any acquirer.

Library of Congress Cataloging-in-Publication Data
Names: Guida, Michael, author.
Title: Listening to British nature : wartime, radio, and modern life,
1914–1945 / Michael Guida.
Description: New York, NY : Oxford University Press, [2022] |
Includes bibliographical references and index.
Identifiers: LCCN 2021036255 (print) | LCCN 2021036256 (ebook) |
ISBN 9780190085537 (hardback) | ISBN 9780190085568 (epub) |
ISBN 9780190085551 | ISBN 9780190085575
Subjects: LCSH: Nature sounds—Great Britain—History—20th century. |
Nature sounds—Psychological aspects. | War—Psychological aspects—History—20th century.
Classification: LCC QH510.5 .G85 2022 (print) | LCC QH510.5 (ebook) |
DDC 591.59/4—dc23
LC record available at https://lccn.loc.gov/2021036255
LC ebook record available at https://lccn.loc.gov/2021036256

DOI: 10.1093/oso/9780190085537.001.0001

1 3 5 7 9 8 6 4 2

Printed by Integrated Books International, United States of America

Contents

Introduction

During WWII, the Ministry of Information produced a propaganda film called *Listen to Britain*. Aimed at a home audience, as well as North Americans with whom Britain wanted to cultivate a good impression, it portrayed in pictures and sound the essence of the nation. Viewers were told that Britain in wartime had a 'pulse like a cannon'. The stern narrator spoke of everyday sound impressions that defined the nation at war: 'The evening hymn of the lark, the roar of Spitfires, the dancers in the great ballroom at Blackpool, the clank of machinery and shunting trains.' This sonic assemblage was a carefully constructed articulation of the multiple ideals of British nationhood; in particular of tradition and technology working together. Anchoring the film were scenes of the tranquillity of the natural world—the cherished countryside, the river Thames, the sun setting over the sea. In the opening moments of the film fields of swaying wheat were swept over by the machine melody of a pair of Spitfires.[1]

The film's message was that Britons were working together in spirited unity and this would get them through the difficult days ahead. Women factory workers sang together while the BBC broadcast its culture across the Empire. Radio listeners were joined together by sound, and victory would be celebrated in sound. Framed by the peaceful yet invigorating sounds of the natural world, Britain was presented as complete and ready to take on the future. After all, the nation was imagined to be built on and anchored by its rural traditions as well as its modern achievements. To be convincing this sound portrait of Britain had to show together the potency of these distinctive elements that made the nation what it was, and the title of the film, *Listen to Britain*, encouraged the audience to recognize and pay attention to them.

Listening, as a way of paying attention to the natural environment and understanding its status and meanings, is the subject of this book. It exposes the significance of the sounds of nature in everyday lives in relation to the changing soundscape of the modernizing world. In part this is a history of the sounds of early twentieth-century Britain, but more importantly it is a history of listening to and interpreting those sounds because with this emphasis I have been able to re-think our understanding of which sounds were significant in the social, cultural, and private lives of the people of Britain. By focusing on listening attention, the book reveals the highly interconnected relationship between nature and modern life—that nature's musicality, energy and tranquilities were highly

Listening to British Nature. Michael Guida, Oxford University Press. © Oxford University Press 2022.
DOI: 10.1093/oso/9780190085537.003.0001

valued and drawn close in despite and sometimes because of the surrounding intensity of man-made or machine sounds.

Listening in Britain in the first part of the twentieth century changed dramatically, perhaps most notably with the appearance of the new media sounds of the gramophone player and wireless radio set inside the home. Outdoors and in the workplace, sounds became unpredictable: the arrhythmias of motor vehicles and their horns, the drone of aircraft and air raid sirens. Contemporaries often categorized these kinds of machine sounds as 'noise' and doctors worried that 'brain workers' were becoming anxious and exhausted in the relentless urban bustle.[2] In the factories, the workshop of the world was noisier still and sufficiently so to permanently damage the hearing. In Glasgow, the auditory acuity of boilermakers faded until words from the Sunday pulpit could no longer be heard, and in Lancashire cotton weavers had to develop their own sign-language and lip-reading techniques to communicate while working. For industrialists and workers alike, the noise of the factory may have had a ring of prosperity about it, but sometimes relief from the roar was needed. The only reliable sanctuary from modern noise, the revered 'countryside', was said by the early 1930s to be at risk of defilement by boisterous day-trippers, careless motorists, and creeping suburban development. The age-old winding rhythm of a Surrey village was threatened by the needs of city speeders in this 1935 excerpt from *The Guardian*'s Country Diary:

> There is much talk about removing these ancient walls to 'straighten' the street so that the London traffic may rush through, unchecked in speed, from town to town. Although many of the gardens may be saved, their mystery will be lost, and their peaceful security will give place to dust, noise and speeding Londoners.[3]

Such impressions have been the focus of much scholarship about the sound of twentieth-century modernity, the large part of which examines the social tensions, politics, and aesthetics of city noise (and not the pandemonium of the factories, steel works, and ship yards, a topic overdue for investigation).[4] This preoccupation with noise in recent academic work has established not just the unpleasantness of loud sound, its characterlessness or alien nature, but also its indication of social disorder, protest, incivility, or difference.[5] James Mansell's recent book *The Age of Noise in Britain* identifies noise, gauged by cultural and medical elites, as a primary pathogenic characteristic of early twentieth-century modernity.[6] Inevitably, social class determined who was deemed to be making noise, who would identify this sonic category, and who could and could not tolerate the sound of it. However, Mansell's account of Britain, together with Emily Thompson's landmark study of sound control inside the offices and auditoria of

American cities, take acoustic modernity to be characterized by noise and the efforts to manage it. Yet, the concept of noise precludes the possibility of listening or paying attention because of the inherent and undifferentiated unpleasantness to most ears of sounds of that designation. An understanding of acoustic modernity must go beyond the debates about and management of unwanted sound. Modern life in this period was in my reading indexed and made sense of through the sounds of the natural world, which were heard amid the din and awarded special meanings in daily life. In some respects, the apparent noise and the sensitivities to it gave new context and meaning to the sounds of nature.

My study departs from Emily Thompson's conceptualization of the 'soundscape of modernity' as one shaped and contained within the interiors of new urban architectural spaces.[7] My notion of the soundscape of modernity opens the doors and windows to show how outdoor spaces were sought out for their sonic qualities, that the sounds of nature were brought inside, and that modern sound and listening included and took forward the sounds of nature. This is not a story of the human design of sonic environments to reflect and project modern life, but an elevation of the sounds of nature as central to modern experience in Britain.

A key question taken up by *Listening to British Nature* is this: which sounds were significant to society?[8] Addressing this question leads emphatically to the natural world not because these sounds were necessarily the most important but because so often the commercialized and mediated sounds of music and human speech have been culturally divisive and contested.[9] However, this is not to say that the sounds of nature have united Britons in appreciation, and in any case no sound is perceived and contemplated in isolation. Peter Cusack's work provides evidence that urban dwellers around the world today hear and enjoy a variety of common sounds that might appear to exist in opposition. In his Favourite Sounds research, Cusack shows the remarkable co-existence of sounds: birdsong, train, and other transport sounds ranking equally highly in the lives of people today.[10] Listening has had to account for all kinds of sounds, but it is notable that those of the natural world have been given continued authority even as populations moved into towns and cities. The meanings of natural sounds have changed of course as social contexts have changed and urbanized, but I emphasize in this study that it has been the quietude, stillness, and silences of nature that have framed modern man-made sounds. Those small sounds of nature were not simply a backdrop to new sounds but contained their own value and took on their own modern meanings. Within the quietude was a complex array of subtle sonic stimuli that the ear was attracted to and worked with. The aim in this book is for the first time to understand the soundings of the natural world in relation to modern life, beginning with the resounding traumas of WWI.

This book examines the period of 1914 to 1945 when war was ever present. As the Hull poet Hubert Nicholson wrote, 'the Twenties were post-war[. . .] the Thirties were pre-war'.[11] In the aftermath of WWI a pervasive mood of anxiety settled across Britain, a period Richard Overy has called the Morbid Age. Recovery from the damage of war was undertaken amid fears that centuries of the civilizing process had been undone in the West, that capitalist economic and political chaos were inevitable.[12] The cataclysm of the Western Front is the starting point of this study, for the fighting there was both the primary cultural and sonic event of British society, matched only by the Blitz period, which forms the endpoint. Considered too is the emergence of the BBC's public service broadcasting in the 1920s, which I regard to be the beginning of a new kind of national listening that quickly attracted an audience of millions, and more than thirty million by the beginning of WWII.[13] *Listening to British Nature* shows how the sounds and rhythms of the natural world were listened to, interpreted, and used amid the pressures of modern life tainted by war. It argues that through the chaos of wartime and the long struggle to recover, nature's voices were drawn close to provide everyday security and sustenance. Civilian soldiers in the trenches, radio listeners, ramblers, naturalists, and medical men all have their say. Nature's sonic presences were not obliterated by the noise of war or the hustle of urban life and its machinery, rather they were given new emphasis by the sensorium and drawn upon to complement and provide alternatives to modern modes of living. Sensuous encounters with the natural world, requiring no education and so open to anyone, opened up the limited sensory repertoire of urban experience. Engaging with nature's sonic and sensory experiences was part of being modern—to be modern was to take nature with you into the future.

There is a need to understand nature as something more capacious than the cosy countryside of southern England. I expand the idea of nature to encompass all of the material terrestrial and cosmic realms because Britons at this time were becoming as intrigued by the tiny physics of Ernest Rutherford's atom as they were by the structure of the universe proposed by Albert Einstein. The characters in this book sometimes considered themselves to be part of nature and sometimes not. In the latter case a vision of a pristine earth excluded humans, nature being 'all that was not touched by man, spoilt by man: nature as the lonely places, the wilderness'.[14] However, for others, humans along with everything else were part of a divine assembly and directed by an inherent force usually identified as God.[15] Such a range of concepts for nature means that listening to it can be various and revealing. Having said that, much of this book will consider nature rooted in the earthbound, where the 'rural' and 'modern' have often in scholarship been held in opposition. This unhelpful positioning has been challenged recently by Kirstin Bluemel and Michael McCluskey, who argue that these terms defined 'a vital relationship that came under intense pressure during' the interwar years. They emphasize the dynamism and interplay of these apparently fixed poles of existence, insisting that the rural was by no means simply a site of nostalgic retreat 'divorced from modernity and modernisation'.[16] Peter Mandler has described early twentieth-century Britain as 'post urban' rather than as a rurally nostalgic and deferential society. He points to suburbanization, the motoring boom, and the rambling movement as progressive features of interwar culture that were more likely to take the town or city to the country than the other way around.[17] It is the perspective of the 'townsman' that I survey here, and this point of view often has managed to juggle traditional and modern values simultaneously rather than fixate on one or the other. The 'romantic moderns' of Alexandra Harris's recent book about the oft-ignored place of the natural and traditional world within British Modernism are people like the anthologist John Betjeman and painter John Piper, who continued to fix their attention on the provincial world of old churches and tea-shops and drizzly English beaches in the 1930s. Harris argues that in dwelling on such seemingly un-modern locales writers and artists explored what it meant to be alive at that moment in England and how the future could be.[18] This book takes together ideas of rural and modern as interconnected, not alternatives. It also acknowledges nostalgia as a symptom of modern times in which the pace of change appears ever to increase. But feelings of nostalgia are not necessarily about travelling backwards; as Svetlana Boym has pointed out, the needs of the present determine the fantasies of the past and have a direct effect on the imagination of the future.[19] The ever-moving experience of the natural world has been constructed in constant dialogue with the past and the future—moments of stepping back into realms of the un-modern have been part of a process of securing the day and moving forward.

Survivors of the Great War needed to take a step back, and those who could did so. Second Lieutenant Ford Madox Ford 'got over the nerve-tangle of war' in the 1920s by 'hibernating' in the peace of the Sussex countryside and gently writing through his wartime experiences.[20] Weary and nerve-wracked, Henry Williamson, who had been a lieutenant in the Machine Gun Corps, locked himself away for four years in his Devon cob cottage to write *Tarka the Otter*.[21] These were the ways for some literary men, but more significantly the government took wide-ranging action after the war to establish smallholdings or allotments to resettle in relative seclusion 24,000 ex-servicemen in England and Wales.[22] Still more took it upon themselves to simply set up huts or old railway carriages on common or waste ground as a place of weekend escape with friends and family.[23] In such places Britons could engage with the aliveness of the natural world within the contemporary world. No sound from nature could weave together the past, present, and future prospects for Britons better than the song of a bird. Birdsong was at once familiar and symbolic. 'We can be sure', wrote the ornithologist and policy-maker Max Nicholson in 1936 about his foreboding of another war, 'that in any case nightingales will sing in Surrey every May, and golden orioles will still flute with civilized perfection in German and French spinneys, regardless of human barbarism or of human achievements'.[24] For many in this study, the voices of birds pointed ahead, echoed into the future, and spoke of the ongoing-ness of the world. Listening in the 1914 to 1945 period made fresh sense of the sounds of nature, giving them active presence in everyday lives, because such sounds were relevant and necessary.

Listening to British Nature considers cultural engagement with the natural world beyond the limiting category of 'landscape'.[25] The static geography implied by the term is bound by the ideals of looking at and painting vistas and 'scenery'. But John Constable was not only thinking of the visual appeal of landscape when he wrote that 'there has never been an age, however rude or uncultivated, in which the love of landscape has not in some way been manifested'.[26] A love of landscape came not just from its surfaces but from contact with and immersion in the living sensualities of the natural world in total. From these experiences, including the fleeting, non-grounded encounters that discussions of landscape can overlook, have been constructed the texture of regional and national iden-tities. It is evident that the gentle English 'south country' has been over-used as a shorthand for many types of Britishness, and more usefully Robert Colls has identified the 'true spirit of England' in the early twentieth century in the modest intimacies of the garden, including the urban yard and allotment.[27] The everyday setting of the garden brings with it the rich sensations of nature, sound in par-ticular animating this miniature landscape. This book will show how the well-known songs of common birds could transform the stolidness of scenery into vibrant present-tense reality. It will demonstrate that all ranks of society with ac-cess to all shapes and sizes of the natural world were beckoned to and solicited by its pleasing sounds, silences, and cyclical rhythms. In his ecological philosophy, David Abram has suggested just how open humans can be to the messaging of the natural world when he writes: 'At the most primordial level of sensuous, bodily experience, we find ourselves in an expressive, gesturing landscape, in a world that speaks'.[28]

It should be clear, then, that a study that concentrates on listening to nature brings specific opportunities to assess who is paying attention to their surround-ings, in what circumstances, and to what ends. This concentration on the sense of hearing does more than add texture to historical understanding, it shows the British relationship with the natural world to be unlike that of existing ac-counts.[29] While foregrounding one sense for analytical clarity, I have accom-modated the others at the same time, in common with others who work on the senses in history.[30] However, Maurice Merleau-Ponty's concept of the body as 'the vehicle of being in the world' is a reminder that any one sense when called to attention brings the concordant operation of others towards the common goal of understanding one's surroundings.[31] And it is the organ of the skin, with its many folds, passages and membranes, that unifies all senses in their interactions with the world.[32] The senses cannot but interact.

In the 1920s listening became increasingly modernized as people became critical consumers of new auditory technologies that could solidify onto wax the vibrations of the air and reproduce them again through a loudspeaker. Even the ephemeralities of a bird's voice could be made modern by recording technology and be appreciated and studied in the home or laboratory, fixed and objectified for the future. The sounds of wild nature could be brought indoors and domesticated. Recorded sound, along with the more pervasive sound of radio broadcasting, made active listening a central activity of public life and a key part of civic engagement in culture and politics in the interwar period, Kate Lacey has argued.[33] The BBC's broadcast reach soon meant that millions of middle-class Britons were listening together and sharing those experiences. Though mass broadcasting elicited much debate among cultural arbiters about the standardizing effects of the words and music pouring forth from the wireless set, also acknowledged was the thrill of imaginative transportation when the medium seemed to become transparent or to disappear completely.[34] How would the sounds of nature fare as listening was modernized in these ways?

Birdsong was a key sound heard and appreciated in interwar Britain, and not just because of the popularization of bird-watching that the new ecologically minded ornithology promoted. The chatter of sparrows, the cascading notes of a skylark, and the iconic, liquid, jug-jug of the nightingale were all resonant in culture. The nightingale's reputation in poetry was matched by no other bird, while its migratory status and range—which rarely extended north of the Midlands—was generally forgotten or forgiven. Apart from birdsong, the book explores the idea and constituents of quiet (and the longed for 'peace and quiet'), which was invariably occupied by swirling leaves or the bleat of sheep. Silence too is considered as a complex category of sound that can attend the psychological states of anticipation or fear, as well as commemoration, religious practice, and the imagined cosmic stillness of the universe. The book highlights as well how human rhythms and circadian patterns have meshed with nature's cadences, taking account of Tim Edensor's argument that the cyclical energies of nature are identified by humans to instil a sense of social order, rationality and stability.[35] Nature has often spelt out rhythms perceived to be an important counterpoint to modern pacing and routine.

Natural sounds and their interpretations stem from the geography, politics, and culture of Britain. Though the materials I have worked with are concentrated in England, I bring in substantive primary source voices from Scotland and Wales, and occasionally Ireland. In the contemporaneous writing I have drawn upon, 'England' was taken to mean, variously, England itself, England and Wales,

Great Britain, or the United Kingdom, and it has therefore not always been possible to avoid repeating these elisions.[36] As this study examines listening in an age of burgeoning empire, an 'imperial patriotism' to some extent can be assumed to infuse all classes of society in their responses to sound.[37] If such patriotism centred on, among other things, a love of nation, woven into this love was a passion for Britain's cherished natural heritage and continuing re-enactments of a pastoralism that resided in the imagination more than the memory or in reality. In any case, back-to-the-land mentalities and the tendency to identify a true national essence within the countryside were not especially English or British and were apparent Europe-wide, as Alex Potts, Jeremy Burchardt, and Peter Mandler usefully remind us.[38] In Britain, such mentalities were not unusual in romantic and privileged thinking on both the political left and right but I do not consider them typical of the everyday aspirations represented in this study.[39] This is not an exposition of how nature pointed Britons backwards, but of how it moved them forwards.

Sources, characters, and chapters

A wide range of British voices will be brought into the open here, especially the written accounts of those who have reported direct experience through 'earwitnessing', to use R. Murray Schafer's term.[40] Letters, diaries, memoirs, and autobiographically oriented books have been the focus for investigating feelings about sounds and listening experiences. Literary sources have been referred to for the historically contingent moods and emotions found there. This approach of course excludes those who did not write, though my aim has been to find points of view across the social spectrum in an effort to give more than a hint of a 'people's history' of listening to nature.[41] The cast of characters ranges from Western Front soldiers, radio listeners, ramblers, and day-trippers, to experts or authorities including medical and military men responsible for the treatment of shell shock, broadcasters and critics, and naturalists and ornithologists. Alex Potts has demonstrated better than anyone how all classes enjoyed engaging with the countryside in this period but in their own ways. The urban middle classes from East London on a trip to Epping Forest might distinguish themselves from the vulgarity of the seaside resort goers and from the materialism and mindless decadence of the country house set at the other end of the social spectrum, yet all had their place.[42] Upper-class and landed perspectives are not given much air here when middle-class opinion and experience are available for analysis instead. Although discourses about sound in the USA have often been linked to the politics of racial difference, exposed vividly in the work of Mark M. Smith and more recently Jennifer Lynn Stoever, in Britain up to WWII race was much less of a determining factor than class in the cultural politics of sound and listening hierarchies.[43]

The chapters are organized chronologically across the 1914 to 1945 period. Working with the accounts of citizen soldiers, the first chapter examines the importance of listening in the trench environment of the Western Front to argue that in spite of the disorder and cacophonous noise, and perhaps because of it, soldiers drew out of the air the chiming musicality of birdsong. They found such sounds cleared and purified the defiled atmospheres of the frontline and provided solace and a sense of future continuity that they might be part of. Paul Fussell's still-provocative poetic concept of 'Arcadian recourse' is developed further to connect to the daily realities of survival.[44] The chapter shows how an appreciation of the sounds of the many species of bird living close to soldiers was cultivated at the same time as knowledge of the sounds of exploding weaponry. The sounds of danger and the sounds of relief were learnt together, and both were needed for survival. The familiar rituals of the lark ascending in the morning and the nightingale's liquid notes in the darkness gave order and rhythm to the unpredictable trench scene. Trench birdlife, heard by British ears, was recognized from home and taken to be part of Britain's fighting spirit. In the stasis of the troglodyte world of the trenches, the song of birds in free flight drew eyes upwards. Here men might contemplate escape homeward across the Channel or an imaginative ascendency into the heavens. Those who returned to Britain and many who spent the four years in London and the southeast of England now had a revised auditory sensibility, which would stay with them for years if not a lifetime.

The second chapter reveals how pastoral quietude emanating from the English countryside was used as a primary therapeutic milieu for soldiers and the nation in recovery after WWI. Quiet rest for mental recuperation was central to influential nineteenth century American medical thinking put forward by George Beard and Silas Weir Mitchell, and it was part of Florence Nightingale's nursing regime. These ideas gain authority in the attempt to treat the bewildering symptoms of shell shock, a phenomenon linked at the time to exposure to explosive force. This chapter makes explicit the sonic dimensions of treatment in country house settings for shell-shocked officers and the belief in the need for gentle pre-industrial ways of working and living after the war for rank-and-file soldiers. For the first time, the archives of Enham Village Centre in Hampshire, a model village experiment for ordinary ex-servicemen, have been brought to light to show how country peace and quiet was part of a national ideology not just for soldiers but for all those shocked by world conflict. Quiet as a sonic category has received scant attention in scholarship. Yet, I have found that its perception and its construction say more than simply an absence of meaningful sound. Quiet can be a container within which small sounds can expand and take on significant meanings. When the invention of the Armistice silence brought so many together to pause and remember, even those two minutes were in fact given over to the sound of leaves and pigeons' wings to inhabit. A dead silence could not be sustained and nor could it heal.

Chapter 3 steps away from the soldier's perspective of listening to nature to consider the emergence of BBC broadcasting in 1922. However, the new national listening service was led by a small group of men who had suffered and survived, so the presence of war had not vanished. The sounds of nature in microcosm and macrocosm are shown here to have a fundamental place in the foundation of broadcasting. It is argued that the BBC's leading figure, John Reith, formulated a vision of broadcasting that was underpinned by the idea that the medium of radio and its programme content could give the nation access to sublime earthly and cosmic silences. In a new reading of the well-known live nightingale transmissions that became an annual event every May, the paradox of a silent broadcasting system is explored. Reith, who described the nightingale's song as a silence that the busy world craved, also saw it as symbolic of the culturally unifying force that his public service broadcasting should be, with an emotional appeal to every citizen. More than this, Reith mixed scientific and mystical ideas to articulate his belief that radio waves connected all humans to the perfect stillness of the heavens and the grand scheme of nature, presided over by a silent God. This chapter considers the first five years of broadcasting, a time when the enthusiasms for wireless listening clashed with concerns about its intrusion into everyday domestic life and notions of what national culture should be at mass scale.

The increasingly confident urban masses were often criticized by observers for their crude and insensitive ways, not least in their use of the interwar countryside. The sensory enthusiasm for the popular pursuit of weekend rambling is the topic of Chapter 4. By examining the habits and experiences of walking in the countryside, moors and mountains a deeply sensual public connection to nature is found. Listening is part of this, but so too are touching and other sensory encounters. These senses, which enable proximate and intimate relations with the outdoors, are considered together not only in order to understand how listening relates to experience more widely, but also to account for the bodily impressions of being immersed in nature that rambling diaries reveal. While rambling in one respect was a rejection of the trials of urban and working life, I argue that the weekend excursions that so many Britons participated in are better thought of as an exchange and balancing out of different sensations available in different places. Having said that, it was encounters with heather, streams, and rock formations that set the pulse racing. Rambling was by no means simply a search for peace and quiet, for hilltop contemplation; rather walkers relished the accentuation of body rhythms and the stimulation that came from direct interaction with the natural environment. What this chapter shows too is that some who use the countryside at the weekend do not go there at all to listen to nature. Instead, the acoustics of the natural world are used to contain and shape celebratory singing, movement, and music.

The fifth and final chapter explores the modernization of listening in an age of new sonic media and what happens to nature's sounds when they can be recorded, studied, and transmitted across regional and national borders. The home front of WWII is the setting, particularly the empire's capital, London, where the broadcasting of recordings of British birdsong is sanctioned throughout the Blitz period and beyond. Here the German nature sound recordist and wartime broadcaster, Ludwig Koch, organized and ordered the sounds of nature and then brought them indoors for millions of adults and children to hear. The sound of birdsong on the airwaves was not tainted by the cultural divisions that human music and speech could provoke in listeners. The potency of birdsong during wartime is demonstrated when senior establishment figures—naturalists like Max Nicholson and Julian Huxley, and the Director General of the BBC Frederick Ogilvie—ruminate about the possibility of appeasing Hitler by broadcasting the song of a nightingale to Germany. I argue that these men were inspired by a vision of civilization strengthened by the presence of birds, whose voices they believed could express British moral values better than human ones. Though birdsong could be modernized by mediation, it did not lose its power.

The BBC is still today a forum for the contemplation of birds and their songs in human lives. Since 2013 BBC Radio 4 has been broadcasting 'Tweet of the Day', a 90-second fragment of a bird's voice and a presenter's voice, just before the grit of the first news programme of the day begins at 6am. In these times of a diminishing natural world, the sounds of birds are heard much less frequently and in consequence may become either more precious or perhaps increasingly irrelevant to human ears. During the COVID-19 pandemic, when this book was being finished, many reported hearing birds calling out in the hushed world of lockdown. The sound seems to have signified sorrow for what has too often been overlooked and lost in the natural world, as well as giving a signal of hope for better days. While the meanings of birdsong continue to change, it is hard to imagine as we listen to birds that human utterance was not first brought to life through the mimicry of animal sounds. This companionship in sound-making, in order to communicate, traces out fundamental continuities of being together with other species in the world. As Edward Thomas put it in his sonnet 'February Afternoon', written in 1916 while he was doing artillery training in England before doing out to the Front: 'Men heard this roar of parleying starlings . . . / A thousand years ago even as now'.[45] Thomas was thinking about the continuity of nature's sound through the ages and how it had become modern with us, becoming relevant in the present. This book intends to show how listening has brought nature closer, how we have been modern *with* nature, modern because of nature.

1

Birdsong over the trenches

The sound of survival and escape

Very rarely are the tongues of Nature stilled. When they are entirely
silent a nameless kind of oppression overtakes the human soul.
Richard Kearton, *At Home with Wild Nature* (1922)[1]

Dead men are hanging in the trees.
Erich Maria Remarque, *All Quiet on the Western Front* (1929)[2]

Trench warfare in France and Belgium in the First World War created a physical and psychic environment that had never been experienced before. No one,
certainly not the civilian soldiers who made up the core of the fighting forces,
was prepared for combat in such trying conditions. Of all the terrible sense
impressions of trench life, the intensity of shelling and machine-gun fire and
their effects on minds and bodies were the ones most written about. The violence of explosive shock waves from bombardments was used to demoralize men
as well as to destroy them.[3] Apart from the gift of luck, physical and emotional
survival on the Western Front has often been attributed to a combination of individual fortitude, soldierly companionship, and communication with loved
ones at home. In addition to these things, this chapter reveals the importance of
contact with nature during life at the Front. Though the fields and woods were
quickly reshaped and broken down into formlessness, something of the natural
world still remained. The aim here is to make clear how listening to nature, when
its coherence was in constant and violent flux, was not only possible, but necessary. Listening to nature's sounds was a technique of solace and survival.

Birds and their songs were the pre-eminent experience and metaphor of hope
in the trenches because they could spur the imagination away from the trou-
bled human world.[4] There was little else in the way of sonic relief to be had in
the trenches or behind the line. Gramophone music was available to a few and
there were concerts and performances put on by professional entertainers like
Lena Ashwell and Seymour Hicks, but these rarely appear in soldiers' writing.[5]
While we know something about the importance of the development of sonic

Listening to British Nature. Michael Guida, Oxford University Press. © Oxford University Press 2022.
DOI: 10.1093/oso/9780190085537.003.0002

knowledge of airborne ordnance, we know little of the wider ways in which listening was deployed in the trenches as a method of survival. Paul Fussell, in
his landmark account *The Great War and Modern Memory*, began to open up
in the chapter called 'Arcadian Recourses' ideas about the role of nature in soldierly experience that have not been followed up by others. He suggested in 1975
that a 'recourse to the pastoral' was an English mode of both 'fully gauging the
calamities of the Great War and imaginatively protecting oneself against them'.[6]
However, Fussell's devotion to the symbolic value of pastoral themes in soldiers'
literary writing means his insight does not sufficiently illuminate the lived experience of warfare. It is the immediate and present-tense responses related to
survival I seek to explore here. Indeed, I have found that while recourse to the
pastoral had its moments, engaging with the vibrancy of the natural world as it
was witnessed and projecting that energy into the future was what preoccupied
men most.

A wide array of sources has been consulted, including letters, diaries, memoirs,
and, to a lesser extent, poetry, written by middle-class civilian soldiers.[7] Letters
are considered as 'field' information, words not intended for publication but to
communicate feelings to colleagues, friends, and loved ones. Although they were
subject to censorship, letters held little back; they provide emotional impressions
of trench life.[8] Three well-known collections of letters were consulted that largely
gather material from men who died on the Front.[9] These edited collections have
been widely used by other scholars and I have concentrated on these published
materials not just because they are accessible, but crucially because they are part
of the canon of WWI letters that has not yet been thoroughly explored for evidence of men's engagement with nature.[10] I have consulted a small set of diaries
to complement the soldiers' letters; those of Edward Thomas, Arthur Graeme
West, Edwin Campion Vaughan, and Wilfred Kerr (a Canadian).[11] Other key
sources include Edmund Blunden's memoir *Undertones of War* (1928), which
I treat as an officer's psychological account as much as a literary-historical one,
and the unusual book *Birds and the War* (1919), which is about all aspects of
bird-life observed in the trenches, gleaned from newspapers and journals and
compiled by Scottish ornithologist Hugh Gladstone.[12] This material allows broad
questions of trench life to be addressed: what did the war of the trenches feel like
and how did men endure the experience? As trench warfare placed new demands
on the senses and the raised the need for auditory knowledge, in what ways was
the natural world in which men were embedded heard in new ways? Beyond the
noise, what *was* heard and what were the realities and fantasies of meaning associated with these sounds and rhythms? In addressing these questions, I have
found Eric Leed's analytical categories of reality, fantasy, and ritual productive
in understanding listening practices in relation to the inner world of the trench
soldier.[13]

First will be considered the modes of trench listening for survival and the presence of birdsong within this sound world. Then, there will be an analysis of three relationships that soldiers establish with nature's sounds; one is a relationship based on sensing rhythms of regeneration, another about perceptions of patriotic resilience, and finally there is the matter of imaginative flight and return home.

'The air is loud with death': Listening in fear for danger

The sound of shelling was the defining sensory provocation of the industrialized warfare conducted across the trenches of the Western Front.[14] Much writing about trench life is preoccupied with the gruelling persistence and noise of bombardment. The intensity of the guns created a steady state of stress and vigilance while shells fell. When they stopped, anticipation of the next barrage was almost as trying. Such sensory assaults had inevitable psychological consequences, the phenomenon called shell shock being closely associated with prolonged exposure to artillery fire. Bayonets, rifle fire, grenade and mortar attacks, and the machine gun made famous in the first days of the Somme were all part of the weaponry of this war. Yet the medieval technology of cannonry, made more sophisticated and powerful than in any previous conflict, was the force that created more fear, carnage, and death than any other mode of killing.[15]

The intensity of artillery action could become close to unbearable. Lieutenant Robert Pickering struggled to cope, and knew that for his infantry embedded in the trenches it was worse still:

> The shell fire never ceases and at intervals regular bombardments take place for hours on end—to put the wind up the other party. This morning early we were in an absolute inferno for a long time and we get that kind of thing about every other day. You get perhaps 5–700 guns going on both sides together, and the number of shells of all calibres that come over are numbered in tens of thousands. It's remarkable how one can live through such an inferno. It nearly drives you mad. Conditions are getting worse and worse for the poor infantry who man the trenches—it is simply an artillery duel.[16]

Under such pressure of attack from the sky, officers and their ranks were left feeling defenceless and exposed. Loud noise itself created acute primal fear, but the results of shelling on the fragile body were equally terrifying. Captain Ivar Campbell, in a letter home in the winter of 1915, told of his bewilderment at the vast quantities of explosives sent in all directions: 'Such infernally large explosive shells to kill such infernally small and feeble animals'.[17] Most of all, artillery was hated and feared for its effect on the body. 'It is all very well to talk of a clean death in battle, but it's not a clean death that the artillery deals. It means arms and legs torn off and men mangled out of recognition by their great hulking bullies of guns,' wrote Lieutenant Arthur Heath.[18] Soldiers crouched, sick with fear, 'not of being hit, but of seeing other people torn, in the way the high-explosive tears'.[19] H. H. Munro, the short story writer and satirist, wrote to his sister Ethel regularly when he was serving at the Front. On Christmas day 1915 he included this adaptation of a carol with a bitter and ironic drawing to explain how the lambs were being blown to pieces in this once pastoral landscape of the Front:

> While Shepherds watched their flocks by night
> All seated on the ground
> A high-explosive shell came down
> And mutton rained around[20]

So often, the sound of artillery bombardment was described as a 'continuous roar'.[21] Ernest Nottingham wrote of 'hour on hour's ceaseless rolling reverberation!'[22] Such descriptions reveal the feeling that the oppression of the guns was ever-present, even though there were pauses and much waiting and boredom in the trenches. These feelings of ceaseless continuity of sound came from the solidity and weight of the explosive torrent in the air. The British fourteen-inch gun could fire a shell weighing 1,400 pounds over many miles, while the 42 cm heavy howitzer of the Germans, nick-named 'Dicke Bertha', could fire shells weighing over 2,000 pounds at the rate of ten per hour. In the eight days from 24 June 1916 of the infamous Somme encounter, 1,732,873 shells were fired by the British. Prior to the Messines assault, early in the summer of 1917, British artillery fired more than three and a half million shells in support of the attack, which equated to at least three shells per second for a 12-day period.[23] A 'storm of steel' indeed, as Ernst Jünger described it in his 1920 account.[24]

The shock waves from explosions would physically attack the ear and invade the body. Ford Madox Ford narrates the opening page of *No More Parades* with these words: 'The drums of the ears were pressed inwards, solid noise showered about the universe, enormous echoes pushed these men to the right, to the left, or down towards the table.' Men were squashed by the acoustic pressure of the guns. Ford continues by saying that even sheltering in a dug-out could not protect men from the noise that 'said things of an intolerable intimacy'. Ford's fiction was deeply rooted in his own frontline experience.[25] A more prosaic indication of the sound pulses tearing across the battlefront comes from William Dyson who, after having had trouble shaving, told his brother that 'the big guns seemed to blow one's beard about'.[26] However, to be close to an exploding shell had its own sonic physicality that was immediate and traumatic. The 'concussion finally made one sick and dizzy', Colwyn Philipps wrote in his diary.[27] New recruits who had not yet got used to the noise would be sent 'green and throwing up'.[28] The physicality was not just oppressive and bewildering; the kinetic energy from a nearby high-explosive detonation could knock soldiers off their feet, stun men into unconsciousness, even stop the heart.

In spite of the apparent wall of sound, the Front was in fact a place of intense and careful listening, in which the differentiation of overhead sounds became vital to survival. With limited vision from a trench position, further obscured by smoke, wire, and mounds of earth, 'hearing became much more important than vision as an index of what was real and threatening'.[29] Eric Leed was one of the first scholars to develop the idea of what Yaron Jean has more recently called 'sonic mindedness' in soldiers fighting on the Western Front, a method of distinguishing between safety and danger.[30] This kind of listening was a military protocol as well as an instinctive individual response. But it took time to learn.

In what he called his 'trench education', nineteen-year-old Edmund Blunden learnt in the first month or so at the Front how to distinguish between sounds that were life-threatening and those that were merely annoying.[31] He found there to be 'a hypocritical tunelessness about a gas shelling in flight and in explosion'.[32] One warm and relatively quiet afternoon in 1915 he tells: 'I called for the company barber and sat meekly under his respectful hand, noting the distance of any disturbances.'[33] Sometimes the rhythm of machine guns from the German side or his own would trace out humour, such as ' "Ri-tiddley-i-ti . . . Pom POM", done in bullets'. From his luminous memoir, *Undertones of War* (1928), it is plain that Blunden and his men were listening constantly for danger and for safety. The Canadian Wilfred Kerr gave his version of the importance of understanding the sound signatures of airborne ordnance in his memoir *Shrieks and Crashes* (1929): 'One was able to judge where a shell would fall by the pitch of its shriek. A noise like a train overhead meant a destination miles in the rear; a sharp shriek or a deep growl meant imminent danger; and between the two there lay a wide variety of pitches, which one soon learned to interpret properly.'[34]

But even seasoned soldiers experienced sounds they had never heard before as technology developed throughout the conflict.[35] And horrible surprises could appear with no warning. In the Cambrin sector of the trenches, Blunden witnessed pure horror as a shell dropped unannounced and turned a young and cheerful lance-corporal making tea to 'goblets of blackening flesh'.[36] Sometimes men were simply too exhausted to pay sufficient attention to the myriad sounds around them: in Wilfred Owen's *Dulce et Decorum est* the trudging troops are 'drunk with fatigue', 'deaf even to the hoots / Of gas-shells dropping softly behind'.[37] Sometimes, knowledge of the components of the soundscape could be useless as sounds lost their definition in the onslaught. 'In a bombardment all tones mingle', Ivar Campbell wrote in a letter from France during the winter of

1915.[38] And for all the knowledge about the kind of projectile approaching, if it was judged to be a heavy shell heading in your direction, sometimes there was little that could be done to protect oneself. It was a matter of 'waiting and wondering', wrote Canadian Lieutenant J. S. Williams, a bank clerk:

> But the real 'corkers' are those 'Jack Johnsons', or 'Coal Boxes'. You hear the brute coming a long way off with the noise of an express train. It's no good hiding anywhere, because you would only be buried by the debris, so you sit tight, hold your breath and pray to God it won't hit you.[39]

Often, then, the sonic chaos of the trenches would have little determinate identity, and however sonic-minded a soldier had become, evasive action was limited.[40] As a result, the trenches were an environment of constant alertness and agitation. It was an environment where ears were always cocked, even if one could not necessarily save oneself. Blunden described as 'mental torture' the concentration of listening to projectiles in flight and detonation over and over again in Bodmin Copse.[41] In this psychic state, men might hear all sorts of things that may or may not have been there. Listening continued into the night, when fear and anxiety were again in attendance. From the position of a listening post poking out into no-man's land, all sounds could conjure dread in the tension and stillness: the 'rustling of grasses', the 'tap-tapping of distant workers', 'the wail of the exploded bomb and the animal cries of wounded men' were all threatening.[42] Soldier-poet Frederick Harvey told of the tension tearing at the nerves during listening-post duty:

> For four dead hours
> Afraid to move or whisper, cough or sneeze,
> Waiting in wonder whether 'twas the breeze
> Moved in the grass, shaking the frozen flowers
> Just then.[43]

Sometimes what sounded like the creeping of an enemy soldier turned out to be a host of rats feeding on the corpses of the unburied.[44]

Some listening activity was more systematic. Artillery units on both sides of the line used mathematical sound-ranging techniques to locate enemy batteries. Complex arrays of microphones were used to distinguish between the sound made by the firing of the gun and the sonic boom of the shell.[45] For all that, the ear might simply be pressed to the ground in a much more intimate connection of the body with the earth, in an attempt to quickly work out the distance of an enemy gun position (Figure 1.1).

Figure 1.1 A distant shell-burst on Pilckem Ridge in Belgium, with a soldier trying to establish the distance of the shelling positions by listening to the ground.
(© Imperial War Museum, Q2739. Photographer: Ernest Brooks)

Underground, particularly from 1916, tunnellers were listening out to determine how close they were to the enemy lines and their own tunnelling works. Miners were enlisted from the colliery towns of Britain to create tunnel systems that could be charged with explosives and detonated under German-held territory, as well as to disable German tunnelling operations. Using special microphone devices connected to a stethoscope, sappers could determine the distance and direction of the picking and shovelling sounds of German mining activity, or even the enemy walking or talking.[46] On occasions, the naked ear had to be put into action. The bellows air-feed would be turned off and all work would stop. Lieutenant Geoffrey Cassels of the 175 Tunnelling Company describes these lonely moments underground:

> Forehead pressed to the face . . . of the gallery, one stood, knelt or lay—listening, listening, listening. Some sounds would be heard, dull and muffled. There was always that fraction of a second of doubt—when it *might* be the enemy mining. One's pulse rate would quicken and fright push to the fore in one's whole being.[47]

For all this carefully considered listening, the noise and intensity of shelling and the destruction of companions and the landscape created an intensely fearful psychological condition that no amount of knowledge of ballistic sounds could fully relieve. But relief was needed and sought out. Amid the bombing there could be respite.

Sonic relief amid the shelling

So much strained listening throughout the day and night had the effect of pulling into the orbit of perception the necessary sounds of counterpoint and relief. In addition, the pummelling of artillery, with its heavy physicality, stimulated a search for lightness, the freedom of the air. The cacophonous and unpredictable soundscape of battle created a need for sounds that felt harmonious, ordered, familiar, and most of all peaceable and non-violent. Men may have become used to the strain of the battle scene to some extent, but they had to remain porous and sensitive to external stimuli. They had to remain alert. Lieutenant Siegfried Sassoon found that trench existence was 'saturated by the external senses; and although our actions were domineered by military discipline, our animal instincts were always uppermost.'[48] One sound in particular appealed to men's animal instincts and became a crucial counter to the battlefield soundtrack—the song of birds.

It was a surprise to men in the trenches to find that birds were even around, that they could co-exist with the shells and the destruction. Captain Medlicott, a man who knew his birds from home, saw and recorded 106 species in Pas-de-Calais in the spring and summer of 1917 between no-man's land and the reserve trenches.[49] Because of the constrictions of the trench system and because of avian behaviour, birds were more often heard than seen. In fact, the two birds that received the most attention from soldiers were the most invisible. The skylark would trill far above the fighting, lost in the sky, and the nightingale's shyness in undergrowth, and tendency to give evening performances, guaranteed its elusiveness. While some soldiers' writing gives the sense that the shelling never stopped, others show that waiting and tedium were frequent. This was a time when birds could be noticed and appreciated. Soldiers were amazed and enchanted by their presence. 'If it weren't for the birds', a Scottish miner turned soldier told his local newspaper, 'what a hell it would be.'[50]

While we have far fewer accounts from men in the ranks than from educated officers, it is just as likely for the former to have been interested and enlivened by seeing and hearing birds on the Front. After all, many of the officer-class were urbanites, whereas their men were much more likely to come from rural and farming families with close affinities with the countryside and its rewards, as well as its trials. The sound of the lark over meadowland was in the blood of country-dwellers, in their folksongs and lore.

The unlikeliness of the survival of birds in trench conditions provoked soldiers to place bets on their fortune.[51] In light of the increasing intensity of artillery action as the war progressed, it became a wonder that there were any signs of the living natural world at all. Much writing from the Front emphasized the obliteration of the landscape, as did Paul Nash's paintings *Wire* and *We are Making a New World*, depicting the headless trees and churned earth of the trench-world. In these images, made in 1918 before he became an Official War Artist, there is a stillness, an absolute absence of life.[52] A similar wasteland was visualized two years earlier by the gunner Percy Smith, a printmaker and artist before the war. He made a dry point engraving of a view at Thiepval, near the Somme, in 1916 (Figure 1.2). The bleakness of the scene is shocking. There is no colour. The lone tree looks like a corpse pointing to heaven. It may be the last tree in the world. What is left provides not even a branch on which a bird may perch or nest or sing from.

Yet there was life in nature still. Theodore Wilson wrote to his aunt in April 1916: 'I'm writing in a trench not very far from the Germans and I've just heard the first cuckoo!'[53] This report from France followed the tradition in *The Times* of publishing claims from those who believed they were the first to hear the herald of spring. It came in the same letter that Wilson announced the gruelling silence of death was everywhere—'you have to walk through it, and under it and past it'—and in this context the cuckoo's simple two-note descending phrase must have sounded all the more striking. The contrast between ever-present death and the vibrancy of birdsong is evident in a poem which William Noel Hodgson formed in his head as he marched back to his rest billets with his battalion after fierce night-time fighting at Loos. The chatter of sparrows is placed in moral opposition to the noise and threat of shelling:

> The foolish noise of sparrows
> And starlings in a wood –
> After the grime of battle
> We know that these are good.
> Death whining down from Heaven,
> Death roaring from the ground,
> Death stinking in the nostril,
> Death shrill in every sound.[54]

After months of fighting on the Somme, with his battalion depleted by three-quarters of its original strength, Edmund Blunden had 'two views of the universe: the glue-ridden formless mortifying wilderness of the crater zone above, and below, fusty, clay-smeared, candle-lit wooden galleries, where the dead lay

Figure 1.2 Percy Smith's 'Solitude' from *Sixteen Drypoints and Etchings. A Record of the Great War* (1930).
(© The Trustees of the British Museum)

decomposing under knocked-in entrances'.[55] In such grim circumstances, it is no surprise to find accounts of the thrill of a bird in song, not just after a bombardment but even during it. Ford Madox Ford, in a letter to his friend Lucy Masterman, spelt out the surprise when he declared 'the noise of the bombardment is continuous—so continuous that one gets used to it', and yet 'the ear picks out the singing of the innumerable larks'.[56] In a state of acute listening, searching for some kind of relief, Ford found his senses could extract from the din the cascades of a much smaller but equally urgent sound. For all their horror the trenches were still able to sustain and encourage bird life, and soldiers grasped the signals ringing out into the battle zone. There was much simple pleasure to be had hearing larks singing through the noise of shelling; 'it is very cheering to see them', Second Lieutenant Otto Murray-Dixon reported.[57] Denis Barnett, a sub-altern in the Leinsters, wrote to his mother from the trenches in 1915: 'it is lovely sitting in the sun, listening to the cock-chaffinches and yellow-hammers tuning up'.[58] For one soldier, the diversion of finding where a golden oriole was singing, even with rifle bullets cracking into the trees above, could refresh the mind. 'For a time', he said, 'the war is forgotten'.[59]

Not for long, though, could the pressures of fighting be forgotten. Instead, a routine of listening was cultivated which accommodated both the sounds of artillery and the sounds of birds. These were the primary markers of danger and hope, opposites on a spectrum of optimism. Each sound accentuated the meaning of the other. One might say that these were the two 'keynote' sounds of the trenches. R. Murray Schafer has argued that such sounds give meaning to all other sounds and have a 'pervasive influence on our behaviour and moods'.[60] These two archetypal sounds of the trenches, birds and bombs, existed in psy-chic tension, one good, one bad, but that contrast also established a rhythm to the trench experience. Birdsong made more tolerable the inevitable sound of bombing.

One artillery officer's diary demonstrates how this exchange of sounds could work to create a rhythmicity to the day. Poet and writer Edward Thomas kept a war diary from 1 January to 8 April 1917.[61] In it, he continued his lifelong habit of observing and recording his response to the natural world. However, unlike his poetry, his diary observations are recorded in sparse, flat notes that seem to reflect his soldierly duties as an officer in the Royal Artillery. There is some irony that Thomas enrolled to be an artilleryman. He was a writer who published in 1909 *The South Country* and then in 1913 *In Pursuit of Spring*, both heartfelt ac-counts of his wanderings and cycling through his precious English countryside. He is the poet who is well-known to have told his friend when asked what he was fighting for to have stopped, picked up a pinch of earth and said 'literally, for this', as he crumbled it between finger and thumb.[62] None of that sentiment is present in his first diary entry from his training station in Lydd in Kent. Rather, it is all very matter-of-fact: 'Shooting with 15 pounders and then 6" howitzers' is his first sentence. Thomas appears content enough and does not mention the noise in these early days of training on home turf.[63]

In France, Thomas' diary entry from 14 March 1917 says more about the sounds around him: 'A still evening—blackbirds singing far off—a spatter of our machine guns—the spit of one enemy bullet—a little rain—no winds—only far-off artillery'. These are notes exclusively about the sound of guns and na-ture, together with weather conditions that would concern an artillery officer as well as someone steeped in sensing one's environment. The seemingly taken-for-granted co-existence of the sounds of mechanical fire and birdsong is again present on 25 March when Thomas writes, 'the O.P. [Observation Post] 20 yards away had a shell on to it, and we had several over our shoulders. Larks singing. Drawing panoramas'. Birdsong also marks out and occupies the quiet when the guns pause. On 28 March, the entry reads: 'Frosty and clear and some blackbirds singing at Agny Chateau in the quiet of the exhausted battery.'[64] He gives little away about how it feels to him to hear birds singing out. There is some pleasure, though his recordings of sound reflect too the monotony of the days, slowly plodding on. Thomas writes short entries every single day in France and almost every one records a sound of warfare or a sound of a bird, or both. His consistent daily writing is a reassuring rhythm in itself, one imagines.

Thomas sometimes enjoyed conjoining the sounds of weaponry with nature's monsters: 'Machine gun bullets snaking along—hissing like little wormy serpents.'[65] On other occasions the descriptions of weaponry take on those of gentle nature, as Thomas becomes accustomed to his daily soundscape. For example, he records on 5 March the 'singing of Field shells and snuffling of 6"'.[66] Does Thomas seek to nullify the threat of the shells by wrapping them in sweet, soft sonic metaphors here?[67] His experience more generally is woven through with observations of weather, moles digging, sycamore leaves dancing. Thomas would also compare the procession of German shells overhead with the sight of familiar birds returning to roost in the evening: 'But Hun shelled chiefly over our heads into Beaurains all night—like starlings returning 20 or 30 a minute.'[68] After a month at the Front Thomas was growing to hate the sounds and vibrations of enemy shelling and needed to find ways to live with them. He wrote of 'a horrible night of bombardment' with little sleep[69] and the 'air flapping all night as with great sails in strong gusty wind'.[70]

Thomas's diary illustrates that hearing the shells he sent over to the enemy and the ones they sent back, together with the sounds of birds, became a defining daily routine that gave shape and meaning to existence. His recording of the beginning of his day on 4 April in three simple steps—waking, hearing birdsong, shooting—brought with it a normalized ritual that could continue indefinitely: 'Up at 4.30. Blackbirds sing at battery at 5.45—shooting at 6.30'.[71] These notes spelt out the components of his surrounding soundscape but also gave a sense of sequence and order to events. On the morning of 9 April, while directing artillery fire, Thomas was killed by the blast from a 5.9-inch shell.[72] The routine was permanently interrupted after just ten weeks in France, when the explosive shock wave stopped his heart. His body remained unmarked. A week before, he had written of the night-time shelling: 'I did not doubt that my heart thumped so that if they had come closer together it might have stopped'.[73] The physicality of sound seems to have killed Thomas, and it has been suggested that the sound waves left their impression on the small diary bound in pigskin that he carried with him—the cover and pages of the diary were left curiously creased with ripple patterns.[74]

Regenerative rhythms

One of the primary crises of trench reality was the challenge of chaos and disorder. This was manifest especially in the unpredictable sound world of weaponry, in which every moment was subject to disruption, shock, and injury. The crisis was also manifest in the potential termination of time for every soldier who saw all around him the ending of life's continuity. In these fracturings of experience can be seen behaviours that sought out a reordering or purification of a corrupted soundscape (where birdsong might clear the air of explosions, screams, and cries) and an active participation in the regenerative rhythms that nature displayed.[75] With these behaviours, trench soldiers were able to ascribe meaning and pattern to their damaged world by mobilizing the cultural resources available to them, even when that world seemed to resist all patterning.[76] All normal daily routines were undermined in the trenches. Periods of frantic activity were followed by longer periods of boredom, waiting, and anxious anticipation. Circadian rhythms were disrupted or reversed as sleep was taken whenever possible, day or night, resulting in disorientation, exhaustion, and depression of morale.[77] In what was certainly an understatement, Richard Donaldson told his mother in a letter in November 1917: 'This is a restless life of ours out here'.[78]

An imbalance of extremes: Noise and silence

Whether shelling was intermittent and unpredictable or the often-recalled 'continuous noise' of intense bombardment, all men longed for some kind of sonic respite provided by the gift of quiet or silence. 'One lives a life of continuous strain and reaction, strain and reaction, and it's difficult to write or think quite quietly', wrote Theodore Wilson, serving with the Sherwood Foresters.[79] This was a war in which men were entrenched in the land and immersed in a perverse rurality. During the mauling and pounding of the countryside by shell-fire, men found themselves searching for the natural state that countryside should be. However, the trench environment hardly allowed for the harmonies of rural sound. Instead, men were caught between fearsome noise and its equally irksome absence. A new day brought hope as the sky glowed once again over a swamp of mud with 'gold and orange and purple fading into an ultra violet', wrote Richard Donaldson, but something was wrong about that scene. 'Only there was no life about it, no men, no birds, no creeping things; and the stillness was painful because the guns were suddenly silent'.[80]

When men wrote of silence, it was rarely dead silence that they heard. The pop of a bullet or the call of a bird as a routine punctuation mark usually appeared within their writing. When silence itself was perceived, death was usually close by. Captain Ivar Campbell demonstrates the unnerving arrhythmias in trench sound, in a description from France on a misty summer morning:

> Save for one or two men—snipers—at the sap-head, the country was deserted. No sign of humanity—a dead land[. . .] there was no sound but a cuckoo in a shell-torn poplar. Then as a rabbit in the early morning comes out to crop the grass, a German stepped over the enemy trench—the only living thing in sight. "I'll take him," says the man near me. And like a rabbit the German falls. And again complete silence and desolation . . . [81]

The night gave special expression to the silence men feared. Anticipation could draw sounds out of the air. On a night mission Alexander Gillespie listened and heard: 'It was a still warm night, and we waited there a long time, expecting to hear the bombs go off. There was a low moon, and a great deal of summer lightning, but it was very quiet, except for a little sniping, and the rustling noises in the long grass.' We do not find out what the rustling noises were, but to record them suggests that there was more in Gillespie's mind than just grass waving in the darkness.[82] Even the much-loved song of the nightingale might disturb, not soothe. Lance Corporal Harold Chapin described to his wife the sensations of visiting a wood where new graves had been placed: 'Over them in the moonlight a nightingale was singing loud and sweet. Its first notes were so close and so low I was startled.'[83] Sweet and yet unnerving were the sounds of nature and the quiet moments they appeared in.

How, then, could a pleasing rhythm to sound events be found? Trees encountered in the night might be silent, solemn friends, or they might appear as phantoms. Sometimes in the 'thick darkness', trees would keep watch over Edmund Blunden and his men. Near Hornby Trench with his signallers he reported the following impression: 'Where we lay, there were in the darkness several tall tree-stumps above, and it felt like a friendly ghost that watched the proceedings'.[84] Even these mutilated trees might act as sentinels in the gloom. If unharmed, trees could be 'noble' and 'a romance and poetry understood by all'.[85] But in the dreaded Thiepval Wood, 'ghostly gallows-trees made no sound nor movement', Blunden reported. Elsewhere, 'one stunted willow' haunted the memory—for Blunden it called to mind Dante's trees in which men and their souls were bound forever into gnarled trunks.[86] On occasion Blunden's trees could be companions even if they were not very talkative—one 'sad guard of trees dripping with the dankness of autumn had nothing to say but sempiternal syllables'.[87] Blunden's ambivalence underlines the impossible balancing of the sonic pattern in the trenches, where threat was possible everywhere, and the meanings of sound were moulded by the moment.

Sensing eternal continuity

Part of the relationship men developed with nature's soundings was closely rooted in the springtime return of new life, and this was heralded by birdsong. 'To-day the frost is all away and we have something like a mild spring day. The birds have been singing since our stand to arms, and various chaffinches and wagtails have come to look at me in my dug-out'.[88] Green shoots appear, orchards continue to blossom close behind the lines, while flowers sprout from incongruous decimation. All these signs of life gain favour and attention and gladden many men, yet birdsong's broadcast energy is perhaps the most powerful proof of hopefulness in the trench landscape. This energy is largely male bravura performance to attract a mate and to command over territory. If it was chiefly bird-minded men who grasped this masculine signalling of sexual determination, many others who heard a blackbird in full voice were taken back to their school-days of egg-collecting and messing about outdoors.[89] Birdsong was remembered from youth, laid down deep in the memory along with all sorts of meanings and associations. Men had grown up with this sound.

Birdsong could encapsulate the continuity of the entirety of nature. This fanfare that seemed to be on behalf of all living things was a reminder that nature, which men were still part of, continued its progress unhindered by human activity. If the tiny specks of bird life could thrive amid the destruction all around, men might stand a chance too. The rhythms and order of birdsong's cadences most pointedly fought against the transience of human life in the trenches because the flow of notes, that some heard as music, continued through time. The notes pointed forward, living on in the mind and propelling the imagination, while shell bursts promised at any moment to banish all future.

The pacing out of the day and night with song is most clearly found in the lark's heralding of the morning and the nightingale the evening. For Ernest Nottingham, the ecstatic song of the lark was 'inseparably connected with "stand to" in the trenches'.[90] To hear the lark an hour before dawn was to know one had survived another night: 'They are wonderful after a night of doubt and terror', wrote Harold Rayner of the Devonshires.[91] To hear the nightingale in the darkness was to be given hope, at least for those moments, that all was well under the stars.

The routines of advance and retirement of troops (three or four days in the line, three in reserve, four or more in the lines behind) created a pattern in which fighting and death would be exchanged for glimpses of the pastoral with all its gentle suggestions of timelessness and growth. Second Lieutenant William Dyson wrote to his brother about the routine behind the line: 'After breakfast rest awhile, walk about the orchards listening to the guns and the birds'.[92] Behind the line, a stillness could be found, one free from fear. Edmund Blunden had a nose for such places. Three kilometres from the Front attending a poison gas course, Blunden found that 'the war allowed a country-rectory quietude and lawny coolness' to pervade, and that the summer could 'multiply his convolvulus, his linnets and butterflies'.[93] Blunden the poet was sensitive to such sensations, but he was surely not the only one.

Nurturing nestlings

There was a particular rhythmic comfort in observing the routines of nesting birds, to-ing and fro-ing in the most astonishing places, often very close to men. Charles Raven was a frontline army chaplain in France in 1915–17. In his book *In Praise of Birds* he recalled an encounter that buoyed up his exhausted battalion. Returning to the line near Vimy, reduced to 150 men of 800, a 'miracle happened':

> Head-quarters was in an old signallers' station, and its entrance was festooned with German wires and decorated with insulators. On one of these a pair of Swallows were building. Those birds were angels in disguise. It is a truism that one touch of nature makes the whole world kin: those blessed birds brought instant relief to the nerves and tempers of the mess. They were utterly fearless, flying in and out among the sand-bags, making the nest ready for its treasures.[94]

Raven continues by telling how the nesting pair were regarded with devoted affection by his battalion, their nest protected while large sums were staked on the date of the first egg's arrival. A trench periscope was used to officially verify the contents of the nest. Raven is delighted at the effect the birds have had on morale.

> No one could be down-hearted when the early 'stand-to' was terminated by the carolling of the cock, and we rushed back to see whether the hen had laid overnight. Blessed birds, they were an allegory of the part which Nature can play for her eldest children when their birthright of toil presses heavily.[95]

There is some sentimentality and of course religiosity in Raven's account, but the witnessing and contact with nature in transformation was a fundamental inspiration. What is important to note is the caring of the soldiers for the nesting birds, their participation in the success of new life, the optimism for a future that was not theirs but reflected a sense of kinship with these small creatures that would soon be independent and fly away. It is not an unusual tale.[96]

The unstoppable progress of nature affects men. It can act as a charm against misfortune or an antidote to the danger that surrounds them. The sonic is a useful way to assess the urgency of nature, because nature sounds as it lives and moves. An example of this, set in contrast with the sounds of a dying man about to become silent, is present in a letter from Captain Wilson of the Sherwood Foresters to his aunt. First, Wilson marvels at the animation brought by spring in 1916, presided over by larks and bees:

> Then a bare field strewn with barbed wire—rusted a sort of Titian Red—out of which a hare came just now, and sat up with fear in his eyes and the sun shining red through his ears. Then the trench. An indescribable mingling of the artificial with the natural. Piled earth with groundsel and great flaming dandelions, and chickweed, and pimpernels, running riot over it. Decayed sandbags, new sandbags, boards, dropped ammunition, empty tins, corrugated iron, a smell of boots, and stagnant water and burnt powder and oil and men, the occasional bang of a rifle, and the click of a bolt, the occasional crack of a bullet coming over, or the wailing diminuendo of a ricochet. And over everything the larks and a blessed bee or two.[97]

But later in the letter, Wilson tells his aunt something of the 'purgatory' that co-exists with spring growth, something he has recently witnessed: 'a bright-eyed fellow suddenly turned into a goggling idiot, with his own brains trickling down into his eyes from under his cap'. This terrible juxtaposition of the skylark's ecstasy and the bumbling of bees with the sound of a man's final moments can only be related to his aunt in England *because* the sounds of nature will endure and remain, as Wilson says, 'over everything'. This is in part how men survive the war and are able to carry on living and fighting—by seeing, but just as powerfully by hearing, nature calling out and announcing its aliveness in spite of the disaster of war. There is some small hope that by association and by inspiration men may be able to carry on and continue in the scheme of life with the birds and bees. Wilson says: 'even the beauty of Spring has something of purgatory about it because all things are seen through a view of obscenity'. But without the spring, what would become of him with only the obscene remaining?

Birds of a timeless universe

Birds were understood somehow to call out ancient rhythms that pre-dated, and would outlast, human activity in the world. We can speak about rhythm in birdsong, as its patterns and repetition are more prominent to human ears than musicality. Its speed and complexity are beguiling yet out of reach. For all that, what is deciphered from birdsong stirs the emotions. An emotional and eternal continuity of the nightingale's song was perceived from Alexander Gillespie's dugout in the small hours of the morning in May 1915. Listening to the song between the bursts of gunfire, he wrote:

> There was something infinitely sweet and sad about it, as if the countryside were singing gently to itself, in the midst of all our noise and confusion and muddy work; so that you felt the nightingale's song was the only real thing which would remain when all the rest was long past and forgotten. It is such an old song too, handed on from nightingale to nightingale through the summer nights of so many innumerable years . . . [98]

As nature's foremost singers and nest-builders, birds could be considered part of the timeless perpetuation of the cosmos. Soldiers in their predicament of the trenches looked for ways in which they could find meaning in events that were not confined to men's struggles. Five days before he was killed in action, Lieutenant Robert Sterling, a Royal Scots Fusilier, wrote to a friend in Glasgow:

> I've been longing for some link with the normal universe detached from the storm [. . .] The enemy had just been shelling our reserve trenches, and a Belgian patrol behind us had been replying, when there fell a few minutes' silence; and I still crouching expectantly in the trench, suddenly saw a pair of thrushes building a nest in a 'bare ruin'd choir' of a tree, only about five yards behind our line. At the same time a lark began to sing in the sky above the German trenches. It seemed almost incredible at the time, but now, whenever I think of those nest-builders and that all but 'sightless song', they seem to repeat in some degree the very essence of the Normal and Unchangeable Universe carrying on unhindered and careless amid the corpses and the bullets and the madness . . . [99]

The indifference of an unchangeable universe carrying on without humans might have been demoralizing, but for Sterling, the thrush and the lark in fact brought him comfort. Nature would proceed and all meaning would not be lost in this war, even if men were to die. Writing after the war, Edward Grey, who served as British Foreign Secretary until December 1916, found security in the beauty and order he heard in birdsong and nature at large. 'Chaos is repulsive', he wrote, and

all people have 'the same impulse to search for law in Nature'.[100] The law that he
and Sterling had pondered was that while men could impede their own future,
the cosmos remained perfect and permanent. By witnessing and accepting these
bird routines as indicative of a greater system, Grey ruminated, humans could
forget themselves and be 'free for a time from moral doubts and strivings'.[101] The
universal rhythms in birdlife allowed humans to give up self-consciousness and
console themselves with a belief in something greater.

Resilience and 'carrying on' in birds and men

'We have a favourite blackbird who sits up in the tree above us, and answers
when the men whistle to him, no matter how heavy the firing may be.' Alexander
Gillespie of the Argyll and Sutherland Highlanders wrote home about his blackbird on 14 March 1915, acknowledging in the same letter heavy losses in his battalion.[102] This blackbird was a kind of pet to the men, a creature to look up to and
converse with. Exposed above the trench works it carried on regardless of the
battle. More than a pet, this blackbird was a friend; that Gillespie reports a dialogue between the soldiers and the bird underlines this relationship.

In assessing some of the letters, diaries, poetry, and writing of the war, perhaps the most apparent sign of hope and survival is that of birdlife carrying on.
Birds had a reputation for toughness and were employed in military operations
because they could accomplish tasks that humans and other animals could not.
Messenger pigeons were a familiar sight to infantrymen. The Pigeon Service
Corps made use of them throughout the Somme offensive, and after that experience some 5,000 birds were added to the service because the birds almost always
got through to deliver their message.[103] Canaries too were part of the combat
scene once the Mines Committee had recommended that two or three birds be
kept at rescue stations to test for carbon monoxide.[104] Tunnelling teams relied on
their sensitivity to poisonous gas and through this partnership canaries gained
admiration for their robustness. A story circulated from a soldier, published
in the Daily Mail and the ornithological journal Bird Notes and News, praising
a hardworking company canary. In all conditions: 'He would do his job under
ground, and as often as not reach the surface again a little limp form lying at
the bottom of his cage; he never failed us though.'[105] These working birds were
known for their spirit. More than that, birds were seen to be *brave*.

Pigeons and canaries were not the birds that appeared in soldiers' writing,
however much their fortitude was respected. It was the activities and singing of
wild birds on the Front that attracted most attention. These birds would inspire
with their apparent resilience and their claim to territory, which often appeared
precarious to say the least. Resilience was what men needed to keep from

becoming victims of the war. To survive, all soldiers had to become hardened, nerveless, able 'to stand without trembling', as infantry sergeant Marc Bloch put it.[106] Singing birds were a reminder of the kind of courage that men would have to foster. Captain Eric Wilkinson of the West Yorkshires dedicated a poem to the strength expressed in birdsong, calling it 'To a Choir of Birds'. This choir, he wrote, was 'insensible to mortal fear'. 'While murk of battle drifts on', the choir was undeterred. He finished on this note: 'The world is stark with blood and hate— but ye— / Sing on! Sing on! In careless ecstasy'.[107] Harold Macmillan noticed the same thing and wrote to tell his mother in June 1916 that the birds sang 'merrily, for all the world as if they were in some peaceful countryside, stranger to High Explosive'.[108] Some sensed a defiance in birdsong, that birds were stimulated to sing out against the guns. Stuart Cloete remembered this in his autobiography decades later: 'There was continuous fire. It was a background to the singing of the nightingales. It has often seemed to me that gunfire makes birds sing, or is it just that the paradox is so great that one never forgets it and always associates the two?'[109] There is affirmation in all this singing in the battlefields that life can continue.

Birds were not expected to be so present on the Front. Ornithologists were themselves surprised to see that birdlife continued busily on the battlefields. A letter in *The Avicultural Magazine* in 1917 told of 'partridges running about between shell craters [. . .], larks singing, magpies all over the place, and a hare lopping along as if nothing were happening, with big guns roaring all round and from every side!'[110] In the same magazine, in the true list-making style of the bird-watcher, a soldier recorded seventy-four different species during his time at the frontline (not as many as Captain Medlicott, who we heard from earlier).[111] It was not just larks and nightingales that were recognized from home. Other experts noted that many of the birds seen and heard were migrants, such as the willow wren, blackcap, and tree pipit, who could have easily travelled further afield to quieter areas. Yet they were present and thriving, as if they were keeping men company.

It has to be said that for some soldiers, singing birds were a jarring insensitivity amid the trials of trench life. H. H. Munro noticed how the skylark had 'stuck tenaciously to the meadows and crop-lands' of the trenches, but he found the 'song of ecstatic jubilation' in the gloom preceding a rainy dawn too much. To him it could sound 'horribly forced and insincere'.[112] The throbbing song of a nightingale in a wood once shattered and reeking of poison gas was an appalling irony to Paul Nash. 'Ridiculous, mad incongruity!' he wrote to his wife; 'One can not think which is more absurd, war or Nature'.[113] Some soldiers took pot-shots at birds because of the unbearable contrast they made in their song and flight with the men's earthbound struggles.[114] 'What the 'ell is 'e singing about?' was the irritation expressed by a prostrate 'British Tommy' when a lark darted into the sky and began its serenade after a day of terrible fighting, according to the *London Mail*.[115]

Nevertheless, such accounts further underline the perceived resilience of birds. There was a Britishness about this behaviour. The determination to 'carry on', as a specific injunction to maintain war-time resilience, had featured prominently in private and public discourses at home during WWI.[116] The idea expressed the desire to 'carry on business as usual', a phrase used by Lloyd George. It referred to the morale-boosting resolve to maintain quintessentially British ways of life whatever the war might bring. To find birds familiar from home singing on the battlefield was in some sense to witness patriotic behaviour. That these were actually foreign birds on foreign ground did not seem to interfere with the perception of birdsong as a display of British patriotic determination. And at home, stories of the heroic birds on the Front circulated in the newspapers. Though birds were behaving in the same ways on both sides of the line, these stories were used to contribute to British propaganda. Birds on the Front were claimed as British.

The iconic nightingale received the most attention and praise, news appearing in *The Times*, *The Scotsman*, and the *Manchester Guardian* of the bird singing in the trenches night and day, in woods stinking of poison gas when men wore respirators and near active gun batteries.[117] The skylark's glorious singing over the trenches, while armies were at death-grips, featured widely in the press during 1916 and 1917. The song of the lark at dawn was 'as usual as the song of the sniper's bullet', proclaimed *The Daily News and Leader*.[118] In the notorious first days of the Somme in July 1916, *The Times* reported the skylark's refusal to quit its habitat and that they could be heard singing during battle 'whenever there was a lull in the almost incessant fire'.[119] Even no man's land proved an attractive place for thousands of birds to nest and rear their young, according to the literary magazine *The English Review*.[120] Britons had chosen to claim that French and Flemish birdlife was singing out and carrying on in support of allied efforts. While birds were understood to be indifferent to the affairs of men, to be living for themselves and singing for their own purposes—careless ecstasy, Eric Wilkinson had called it—they were still claimed as taking sides and championing Britain's efforts.[121]

There must have been a different kind of propaganda at work when this extraordinary tale appeared about British soldiers in France reduced to tears by birds in song. In the *Daily Mail*, next to articles about blinded soldiers returning home as heroes and trade restrictions with the enemy, this small piece nestled:

> The Rev E. L. Watson, an Army chaplain, said that during the bombardment of Neuve Chapelle several men were seen to be crying. When he asked why, they said that during the lulls of the bombardment, birds could be heard singing.

Over one trench hung a tree, and on a bough were two birds preening their feathers and twittering.[122]

Singing birds in this case were presumably used to illustrate the sensitivity of British soldiers. Their tears were indicative of their humanity and their civilization, but at risk in this story were their soldierly qualities of manliness and pluck. If these British soldiers were distinguished from the barbaric and sadistic Hun by their fine feelings, they were at the same time inadvertently depicted as part of an army who would not be successful.[123]

In large part, though, wild birds were taken to be agents of British fighting spirit at its moral best. Pet canaries were given a role too. They had been commonplace in domestic working- and middle-class culture (along with others in the finch family) since the mid-nineteenth century, and much loved by miners, who set up 'fancies' to breed canaries.[124] For generations, canaries had been kept for their cheerful song and perkiness, and in wartime this was seen to be a tonic that would build strength and assist recovery in injured soldiers. Canaries were put to work in ambulance trains to 'cheer our wounded soldiers with their sweet song' (Figure 1.3).[125] At home, Private Dobson's recovery in a Southampton hospital was aided when the hospital arranged for a Yorkshire show-bird to be sent, which sang in the sunlight beside his bed.[126] These stories reveal an intimate link between Britons and birds. The song of a bird often served to lift the human spirit upwards.

Skyward escape with the lark

Perhaps the most important aspect of birdlife on the Front for soldiers was its occupation of the sky. Birdsong was airborne and free. Flight could not but suggest the dream of escape. In no bird did these facets come to life more than in the skylark with its ascending flight and cascading song. The lark was a well-known field and meadow bird in Britain, yet it had been invested with magic in poetry: 'the Lark is a mighty Angel', Blake wrote; a creature from 'heaven, or near it', said Shelley.[127] In the first week of the war, on Margate's cliff top, Ralph Vaughan Williams was writing the musical setting for George Meredith's poem 'Lark Ascending' and the moods he was working with were simultaneously emerging in the trenches.[128] For the Machine Gun Corps Captain, Edward de Stein, the bird was a 'happy sprite'.[129] The skylark was associated with the heavenly vault untouched by the war, its skyward ascent, away from the trenches and corpses, could lift men up and away, at least momentarily. Always hard to see, the sound of the song drew attention to the freshness of the sky which was renewed each morning.[130]

Figure 1.3 Nurses with caged canaries on an ambulance train near Doullens, France, April 1918.
(© Imperial War Museum, Q8739. Photographer: David McLellan)

The earthbound realities of trench experience necessitated alternative imaginings. Escape with the lark would take men away from the stasis of trench life, the immobility. Men had become part of the landscape itself, dug-in, mud-caked, trench-footed, and concealed in the ground like the rats they shared it with. Entrapment in this world under the weight of shelling put men under great psychological strain, which became all the worse when you could hardly move.[131] Writing about the effects of high explosive shell-fire on the mental state of soldiers, the physician David Forsythe remarked that 'experience has shown that a high degree of nervous tension is commonest among men who have, perforce, to remain inactive while being shelled'.[132]

The strangeness of the trench scene begged for relief. Strangest of all was the presence of death all around.[133] Some frontline areas were like graveyards, Rowland Feilding observed: 'Many soldiers lie buried in the parapet, and in some cases their feet project into the trench'.[134] Ghostly remains littered abandoned trenches, 'old uniforms, and a great many bones, like broken bird cages'.[135] David Jones, a private in the Welsh Fusiliers, wrote of 'a night-feast on the broken of us' in his poem 'In Parenthesis'. Of the sounds of dead men being feasted upon he was explicit: 'You can hear the silence of it: / You can hear the rat of no-man's-land / rut-out intricacies, / weasel-out his patient workings, / scrut, scrut, sscrut.'[136] Worse than the sound of the dead was the sound of their dying. There was 're-lief that the thing has ceased to groan: that the bullet or bomb that made the man an animal has now made the animal a corpse', admitted Charles Sorley.[137] Wilfred Owen's unforgettable 'guttering, choking' of a man overcome by gas still shocks. Writing home on 4 February 1917, Owen articulated the grim picture of the earthbound better than anyone. He described:

> the universal pervasion of *Ugliness*. Hideous landscapes, vile noises, foul lan-guage [...] everything unnatural, broken, blasted; the distortion of the dead, whose unburiable bodies sit outside the dug-outs all day, all night, the most execrable sights on earth.[138]

Some kind of escape was longed for. It was found in the camaraderie of men, striving together, telling jokes, drinking tea and never speaking of 'anything de-pressing'.[139] All this 'fellow-feeling' was necessary.[140] But only so much comfort could be expected from humans making a war like this.

It was the lark that turned eyes upwards, away from the trench system. At dawn, when there was usually a stillness intact from the night, the lark's trill would rise, taking men with it. For Ernest Nottingham, though he writes of the 'hardening of experience', the soaring song at stand-to could lift him up and away from the 'bones and barbed wire' that he had seen that night.[141] Returning from night duty, Private Isaac Rosenberg and his patrol were enchanted by a burst of lark song from the early morning gloom. He knew that 'death could drop from the dark' as they trudged nervously back to camp. But the larks formed an invisible shield. In 'Returning we Hear the Larks', he wrote these lines:

> But hark! joy—joy—strange joy.
> Lo! heights of night ringing with unseen larks.
> Music showering our upturned list'ning faces.[142]

As Paul Fussell has argued, the sound of the skylark provided men with 'evi-dence that ecstasy was still an active motif in the universe'.[143] The delight of the lark's song had the potential to draw men's minds out of the material world and into the heights of reverie, a place of dreams and sustenance. John W. Streets, a Derbyshire miner before the war, now with the rank of Sergeant, found the lark could transform his spirits: 'My soul rushed singing to the ether sky'.[144] The trilling of the bird spoke even to the 'long-lost dead' in ways that 'gave the kiss of life', Streets felt, perhaps because the lark sang so close to heaven. These are powerful testimonies, and there are many others, that speak of more than the joy of hearing a lark in song. There is also the desire in these reports, poetic though many are, to rise up and to not look down. These are momentary releases from the mud and mess, the antithesis to the most common soldier's dream of being buried alive in a bunker by a heavy shell.[145]

Captain Colwyn Philipps, a professional Sandhurst soldier posted with the Royal Horse Guards, kept a notebook on the Front. It was found after his death near Ypres in May 1915. There was a poem there that explored how he might rise above the 'earth's stain', beyond the sky and up towards the stars. He felt that the night-time was when this might happen, a time when stillness was in the air. Then, for a short time, he would try to find his sense of God, reaching towards the stars that were suspended above the battlefields, before returning again to the world of war. This connection to the peacefulness of cosmic nature, made in a visit of just 'an hour', was an escape he needed. Philipps's poem, quoted here, is marked by his father, who published his poetry and letters from the Front in 1915, 'Found in his note-book when his kit came home'.[146]

> There is a healing magic in the night,
> The breeze blows cleaner than it did by day,
> Forgot the fever of the fuller light,
> And sorrow sinks insensibly away
> As if some saint a cool white hand did lay
> Upon the brow, and calm the restless brain.
> The moon looks down with pale unpassioned ray—
> Sufficient for the hour is its pain.
> Be still and feel the night that hides away earth's stain.
> Be still and loose the sense of God in you,
> Be still and send your soul into the all,
> The vasty distance where the stars shine blue,
> No longer antlike on the earth to crawl.
> Released from time and sense of great or small,
> Float on the pinions of the Night-Queen's wings;
> Soar till the swift inevitable fall
> Will drag you back into all the world's small things;
> Yet for an hour be one with all escapèd things.

Imaginative transportation home was another mode of escape for the trench mind that larks and other birds would provoke, sometimes with melancholy. Frederick Keeling, a Sergeant Major in the Duke of Cornwall's Light Infantry, found himself plagued by nostalgia:

> Every morning when I was in the front-line trenches I used to hear the larks singing soon after we stood-to about dawn. But those wretched larks made me more sad than almost anything else out here . . . Their sounds are so closely associated in my mind with peaceful summer days in gardens or pleasant landscapes in Blighty.[147]

'Loved to our Gloucester eyes' were the birds that Private Ivor Gurney and his regiment encountered around the trenches. Hedge sparrows, 'laughing linnets', and a yellowhammer were treated as old friends visiting the soldiers from their home county where the men associated them with quite specific rural localities, named in Gurney's poem 'Birds'.[148] It was a surprise and a delight to see and hear 'some visitor from home with a touch of the rhymes'. Home had travelled to the trenches with these visitors, as far as Gurney and his mates were concerned. When Paul Nash spotted a swallow from a hillside, he wrote to his wife: 'Ever since he has haunted me. I can't help thinking he may have been on his way to England where my sweetheart is and you may see him just as I did skimming the fields.'[149] His swallow was a messenger to his wife at home, a metaphor for connection. Men thrilled to the reality of freedom of flight that birds had. This mastery of the air seemed to promote feelings of contact and communication, away from the trench experience.

Birds shared the sky with aircraft, of course. To sight an aircraft was a greater fascination than a bird though. There was a fear of enemy machines, but the airman took on the role of a superior being in comparison to the fighting masses on the ground.[150] The airman was free and worked at a distance from the real war. He could fly with 'a sort of triumphant calm through the tiny snow-white puffs of shrapnel' that would surround him.[151] Skylarks, high up, could be confused at first sight with aircraft.[152] In this muddling of larks with aircraft there is again the romance of flight and the supposed imperviousness of the flier. In one of Edward Thomas' typical juxtapositions of war technology and birdlife he wrote this entry in his diary in February 1917: 'Black-headed buntings talk, rooks caw, lovely white puffs of shrapnel round planes high up.'[153] The scene is rather picturesque. While the birds call, the war planes seem here to be immune to shrapnel fire, which instead makes soft little clouds in the sky.

By assuming an aerial perspective, the soldier at the Front could gain a psychic distance from the crushing actualities of the trench environment. Eric Leed has said that 'the flier, in fact and fantasy, keeps open the possibility of an escape'.[154] A bird or an aircraft could offer a vantage over the scenes that the soldier was participating in and secure his survival in the mind's eye. But a bird was better because it was not involved in the war. Its song was interchangeable with flight, important for soldiers whose vision was always constricted—to hear a bird was to fly. Looking through the eyes of the bird or riding on its back was to oversee, with God-like vision, a purpose in the war. Or perhaps by placing oneself behind the eyes of a bird one could escape the point of view of the human.[155] Or one could soar on the breeze, travel across the Channel and, in no time, find oneself at home among the tenderness of loved ones. Birdsong could whisk the imagination upwards and homewards.

Conclusion

Given the standard accounts of artillery noise and the familiar artistic depictions of a macerated Western Front landscape, it seems surprising, even incredible that men heard birds amid the combat experience of the trenches. Yet, numerous accounts from the Front document both birds and bombs as the two key sounds competing for attention in soldier's minds. The performances of larks, nightingales, blackbirds, sparrows, and cuckoos were written down and sent back home to loved ones to show that some cherished things remained unchanged even in wartime. The sounds of birds were a shared vocabulary of optimism. A question hovers here as to how much could actually be heard amid the cacophonies of shelling. However, it can be assumed that both the 'continuous roar' and the singing have been accentuated in soldiers' accounts, because each were so meaningful. These extraordinary but daily auditory experiences may have been easier to write about than the many visceral horrors that presented themselves. So many aspects of frontline experience had to be allowed to sink 'like stones' the moment they were over.[156] Birdsong, and its companion sound gunfire, stayed with soldiers, emerging in their letters at the time and in memoirs years later.

It was the concentrated listening practices, forming part of everyday survival, that brought to the ear all sounds, great and small, and gave them heightened significance and meaning. As men acquired knowledge of the sounds of weaponry, at the same time they began to know the scratch of rats, the sound of wire being repaired, the shrieks of wounded men trapped in no-man's land, and the songs of birds. Birdsong might have been a practical marker of safety for a moment or two, but it functioned as an emotional and spiritual tonic more than anything. Birdsong came to signify a purity and harmony in the air, a resilience and continuity of spirit, the hope and freedom implicit in all living beings, and a metaphor for taking flight. It could even occupy and refresh the lulls and silences that were otherwise crackling with anticipation and anxiety.

To hear a bird in song was to hear an urgent celebration of life's ringing energy when death was the surrounding motif, spelt out in explosion or in silence. Birdsong was a tiny but piercing broadcast that officers and their ranks drew down from the air, though they knew this communication was not intended for them. Men felt they had a kind of dialogue with the birds. They were spoken to as well as serenaded. When the war was over, Siegfried Sassoon imagined that all soldiers became birds: 'O, but Everyone / Was a bird; and the song was word-less; the singing will never be done.'[157] The relationship with birds through their song centred on presentness, on existence in that moment when a bird's song was heard. Birds were not heard every day by any means, yet birdsong might inter-vene in many of the everyday functions on the Front; those of waiting, staying connected with family and friends, camaraderie, and managing fear.[158]

Hearing, as a primarily temporal and ephemeral sense in which sound waves operate through time, is subject to interpretations of rhythm and dis-rhythm, continuity and discontinuity. This analysis has found that birdsong marked out time in a way that emphasized the rhythms and progression of the natural world. Its sound did not just fade immediately away. Rather its resonance seemed to live on, ascribing a pattern to the world. For many officers and ranks the lark heralded the morning stand-to, the nightingale the evening one. These sounds signalled daily renewal. But the reassurance of seasonal renewal too had a marked effect on men who might see and hear several seasons pass during their service. With the springtime reappearance of birdsong in the air came the reminder that even in war, the universe was still able to sustain fragile new life. Birdsong today brought the promise of tomorrow.

The new kinds of sonic sensibility that were cultivated during the war did not vanish with the sound of the guns. Those who returned to Britain from the Front, and many who spent the four years in London and the southeast of England lis-tening to the guns, now had a revised sensitivity to sound, which would stay with them for years. The next chapter explores how the continued influence and knowledge of sound was part of recuperative thinking and practice for men with shell shock and for the nation in recovery.

2

Pastoral quietude for shell shock and national recovery

> Be still, my soul, and sound
> The stillness all around:
> Brings peace to wearied souls,
> 'Tis Nature's Paradise.
>
> A convalescent soldier, *The Hydra* magazine (1917)[1]

In the summer of 1917, Siegfried Sassoon was recovering in London from a bullet through his throat. He was fragile and in 'a rotten state of nerves'. When he went for a walk he saw corpses lying about on the pavements. In the ward of his convalescence home, he found the gramophone plagued him to the limit of his patience. Sassoon longed for some peace: 'If he could only be quiet and see no one, simply watch the trees dressing up in green and feel the same himself'.[2] There were many soldiers like Sassoon, with physical wounds, mental damage, or both, who needed a ration of quiet rest in sight of nature to be able to continue.

Britain did not fall silent after WWI, but in its aftermath ideas about quiet and silence took on new cultural and medical significance. Sensitivities to sound were certainly heightened, but the meanings and values of sounds had changed too. This chapter reveals how pastoral quietude emanating from the British countryside was used as a primary therapeutic milieu for soldiers and the nation in recovery after WWI. The catastrophe of this war demanded a sense of the pastoral as its antithesis. Paul Fussell made this link when he wrote: 'if the opposite of war is peace, the opposite of experiencing moments of war is proposing moments of pastoral'.[3]

The treatment of shell shock is a focus here, as this condition and the metaphors surrounding it reverberated widely through society during the 1920s and beyond.[4] Though the medical response to shell shock was one of bewildered therapeutic experimentation with a host of techniques, the tradition of quiet rest provided the backbone to almost all interventions for men who had been mentally or physically injured. The best kind of recuperative quiet came from rural settings, a notion that came from American medical trends along with Florence

Listening to British Nature. Michael Guida, Oxford University Press. © Oxford University Press 2022.
DOI: 10.1093/oso/9780190085537.003.0003

Nightingale's influential nursing regimes. Placement of officers in country house environments may seem like a natural enough return to privilege, but rank-and-file soldiers on occasion were also situated within the healing power of the countryside. For the first time, the work of Enham Village Centre, a model village founded in Hampshire in 1919 for ordinary soldiers which still exists today, has been brought to light to show how country peace and quiet was part of a national ideology of recuperation not just for soldiers but for all those shocked by world conflict. National recovery employed symbolic silence in the Armistice ritual, but only briefly, because dead silence could not heal. More pervasive through the interwar period was the sound of new national music, most notably from Ralph Vaughan Williams, who attempted to bring to modern currency pastoral motifs and allusions of identity rooted in the land.

To concentrate on the shell-shocked soldier may risk his further historical overexposure; however it was only in 2002, with Peter Leese's *Shell Shock: Traumatic Neurosis and the British Soldiers of the First World War*, that the first full-length English-language historical monograph on psychological trauma in this conflict was published.[5] Moreover, to concentrate on the phenomenon of shell shock is to acknowledge a potent cultural symbol that quickly escaped medical discourse and became an expression of societal fracture by the war long after the guns had stopped. An immersion in the post-war public debates about shell shock allows the historian to gauge the atmospheric conditions of the times—the psychological weather. The official figures, which suggest 2–4 per cent of all admissions to British military hospitals were psychiatric battle causalities, are likely to be 'wildly inaccurate', Jay Winter has argued, and say nothing of the burden placed on families for many years to come.[6] In the decade following the Armistice about 114,600 men applied for pensions for shell shock-related disorders, and yet such statistics give no impression of the tremendous social and economic impact of the war in homes and workplaces throughout Britain.[7] A mood of anxiety pervaded many lives in the 1920s because the war had revealed the failure of civilization and taken-for-granted socio-economic norms. Could civilization be made to work again? Could romance in life be recovered? Britain was desperate to return to tranquillity and social peace and this was manifest in the emergence of the League of Nations and in the 1930s the production of the Peace Ballot and formation of the Peace Pledge Union.[8]

For all the voluminous literature about shell shock, few scholars have highlighted the use of quiet rest as a foundation for most treatment regimes and convalescence.[9] Brendan Kelly and Fiona Reid have considered the place of quiet in the recovery programmes of shell-shocked men sent home. Kelly's work has drawn attention to the consideration of quiet rest by the 1922 *Report of the War Office Committee of Enquiry into 'Shell-Shock'* and has found, in his research into Dublin's Richmond War Hospital, practices that emphasized the importance of quietude.[10] Reid has examined the use of quiet rest, stemming from the nineteenth-century rest cure, arguing that 'the most obvious and consistent characteristic of government plans for the post-war care of shell-shocked men lay in its commitment to a rural system of treatment'.[11] When the War Office returned shell-shocked officers to the peaceful world of a country house, it sent them back in time to an idealized pre-war England, she points out.[12] This was more than a statement of class privilege; it drew on long-held views of the countryside as the locus of healing. Jay Winter, in his most recent writing about shell shock, concludes his review of the many approaches to treatment with the following statement: 'On balance, doctors did the best they could, which many times was simply to put patients in a quiet environment, where some spent the rest of their lives.'[13]

Tracing the lives of men with shell shock has hardly been possible for historians. Many men who survived would not talk about their experiences of war, and those with shell shock were perhaps the least articulate of all, some being temporarily muted by their trauma.[14] Instead, the voices of medical men will be heard, though many of them speak as officers as well. They have been treating injured men in hospital, designing new recovery programmes, pondering the efficacy of musical therapies or engaged on government committees. What have these voices to say about the ways in which the quietude of British nature was imagined and used across society as an antidote to the psychological shock of WWI? Among the array of far more muscular interventions for shell shock lies the seeming simplicity of quiet rest, which has been largely lost from view.

Quiet for the wounded?

Within the first few months of the conflict, men began to be sent home from frontline service with a perplexing array of symptoms—stupor, paralysis, tremors, psoriasis, delusional states, depression, nightmares, and nervous collapse. Physicians tried all manner of treatments, reflecting the many and often conflicting views about the origin and nature of these disturbances. Only soldiers with the most severe symptoms were taken away from the battle and the priority was to promptly restore them to fighting effectiveness.[15] Initial treatments often sought to produce a disciplinary fix through the use of isolation, restricted diet, or even electric shocks to alter soldiers' apparently aberrant behaviour.[16] Other treatments were geared towards addressing a potential unconscious psychological conflict in the soldier's mind. In this case a military therapist would use hypnosis and the re-experiencing of traumatic memories to purge them of their emotional impact.[17]

However, there were other equally or more effective approaches to the management of shell shock based primarily on the provision of rest in a quiet environment. Such seemingly gentle strategies were most clearly articulated after the war as part of the two-year deliberations of a committee led by Lord Southborough. The highly political 1922 *Report of the War Office Committee of Enquiry into 'Shell-Shock'* was a distillation of clinical and military expertise gained during WWI and its immediate aftermath.[18] Evidence from influential surgeon and psychiatrist Edward Mapother, who had served in France and Mesopotamia and then at Maghull psychiatric hospital near Liverpool for rank-and-file soldiers, was reported as follows: 'Dr Mapother thought every anxiety neurosis case in its very early stage could have been cured if taken out of the line and sent to a rest camp.'[19] In its final recommendations, the committee warned against the indiscriminate use of therapies based on discipline, hypnosis, or re-experiencing because they might aggravate symptoms. Instead, the committee placed a strong emphasis on giving the exhausted nervous system a rest:

> Good results will be obtained in the majority by the simplest forms of psychotherapy, i.e., explanation, persuasion and suggestion, aided by such physical methods as baths, electricity and massage. Rest of mind and body is essential in all cases.[20]

Rest could only have taken place in quiet, comfortable surroundings where the senses had a chance to stand down. It is true that rest and quiet could only ever be relative terms in wartime, the rest huts of the Front provided by voluntary aid organizations being a case in point. Their homely atmosphere came from the chance to enjoy a cup of tea and a biscuit, a newspaper, a gramophone record, a game, a sing-song, or a religious service.[21] Rest huts were in earshot of the fighting and yet they were intended to create a mood of relative quietude that would allow men to relax, sleep, and regain their strength.

In the absence of any effective treatment, shell shock called for the commonsense provision of quiet, not least because the condition was immediately associated with intense, prolonged battle noise and the overwhelming of several senses at once.[22] Charing Cross Hospital physician David Forsythe believed that assault by noise was part of an assemblage of stimuli from a shell blast that could contribute to shell shock:

> The detonation, the flash, the heat of the explosion, the air concussion, the upheaval of the ground, and the acrid suffocating fumes combine in producing a violent assault on practically all the senses simultaneously, and the effect is often immediately intensified by the shrieks and groans and the sight of the dead and injured.[23]

Charles Myers' term, 'shell shock', suggested both the surprise in witnessing a sudden nearby shell explosion and the physical force of the blast wave energy on the body, though he later found that men who had not been subjected to shell fire could display similar symptoms.[24] Nevertheless, for Myers, quiet rest was a logical and humane solution. He would not participate in the Southborough committee enquiry (because his psychological idea that shell shock was an emotional condition that could and should be treated had not been accepted by the military authorities), but the report gladly quoted his 1916 opinion that emphasized the need for a restful sonic environment, so long as it would also serve as a reminder of the duty awaiting the soldier with mild mental distress:

> The centre to which these slighter shell-shock cases are first sent should be as remote from the sounds of warfare as is compatible with the preservation of the 'atmosphere' of the front. It must, therefore, be neither within easy range of bombardment, nor within sight of England.[25]

The Southborough report would not entertain the idea that a retreat towards England could be efficacious because it did not accept the premise of shell shock as a medical condition; rather it was a sign of weakness and probably cowardice. The report's views about where men should be treated went against the guidance from the most influential nursing text of the war.[26] Violetta Thurstan's *A Textbook of War Nursing* asserted that the most important way to deal with shell shock was 'complete mental and physical rest in bed, so that these patients are always sent home from the front as soon as possible, right away from the scene of the war'.[27] The lack of consensus meant that only sometimes did men in psychological crisis find themselves allowed quiet rest. In fact, some military physicians who favoured disciplinary styles of treatment suggested that 'some ordeal of noise' for new recruits would help accustom them to the sound of battle and so prevent shell shock developing.[28]

There were two strands of medical thinking that gave authority to quiet rest as part of the regimen required for recovery from mental breakdown. First was the continued prominence of the nineteenth-century orthodoxy of the 'rest cure' for neurasthenia, an idea that first flourished in the USA. Neurologist Silas Weir Mitchell formalized the notion of the rest cure following the study of paralysis in soldiers during the Civil War. He went on to develop his therapy to help nervous and apparently hysterical well-to-do women. For them the cure entailed several weeks of bed rest, isolation and a rich milky diet.[29] Versions of Mitchell's rest cure were adopted during WWI when British physicians and neurologists thought shell shock might be a kind of neurasthenic condition of the nerves.[30] Secondly, Florence Nightingale formalized the notion of quiet being part of the recovery regimen for wounded soldiers when she established her wards in the

Crimea. Her widely read and influential *Notes on Nursing* dedicated a chapter to the management of noise. Patients with irritable nerves should be surrounded by silence, she counselled, because no amount of fresh air or careful attendance could achieve anything without quiet.[31] Nightingale's strong views about sonic discipline were clearly in evidence and further reinforced in nursing practice during WWI.[32] It may well be that Violetta Thurstan's advice about dispatching men away from the noise of combat reflected Nightingale's earlier theory.

These two little-discussed founding principles and practices of shell shock treatment—quiet and rest—were strongly linked to natural outdoor environments, away from towns and cities, where such conditions could best be found. Mitchell had devised a special version of his rest cure for men, and he called it the 'West cure'. It involved cattle roping, rough riding, hunting, and bonding with other men in rugged outdoor locations. Famous recipients of this nature cure included the future US President Theodore Roosevelt, painter Thomas Eakins, poet Walt Whitman, and novelist Owen Wister.[33] Eakins recorded his experience of the West cure in a series of letters, sketches and paintings (Figure 2.1). His serene depictions of men in the wilderness of the Dakota Badlands show a side of the cure apart from the heroics, where restful retreat and reflection within the landscape could calm the nerves and rebuild the spirit.

Another American, George Beard, who defined the condition of neurasthenia, had argued in 1881 in his book *American Nervousness* that 'the moans and roar

Figure 2.1 'Cowboys in the Badlands' by Thomas Eakins (1888).

of the wind, the rustling and trembling of the leaves and swaying of the branches, the roar of the sea and of waterfalls, the singing of birds, and even the cries of some wild animals' could be considered therapeutic for their rhythmical and melodious character.[34] These raw sounds of nature, even savage ones or those that might evoke a sense of the sublime, were distinct from the harsh and ar-rhythmic sounds of modern life. Beard's ideas about the nervousness of 'modern civilisation' were taken up in Britain and complemented existing thinking of how to look after the mentally ill.[35] In Britain, health benefits were attached to the sa-lubrity of outdoor places where fresh air could be had and sunshine could reach the cheeks. Implicit in the fresh air doctrine was prolonged exposure to a natural soundscape. Those responsible for therapeutic management in lunatic asylums and sanatoria emphasized the importance of tranquillity inside as well as in the grounds and gardens to create a mood of cheerfulness and healing.[36] Samuel Tuke's York Retreat for the mentally ill, run on Quaker principles of pastoral care, was carefully depicted in early nineteenth-century engravings as a bucolic haven of peacefulness, nestling in leafy unspoilt countryside.[37] The humane 'moral therapy' that Tuke pioneered relied in part on the wholesomeness of the English countryside in which fragile minds could be nurtured.

What follows are two significant examples of the use of pastoral quiet during wartime and its aftermath, first in country house settings and second in a thera-peutic village community.

Country house therapy

The British public wanted to see that all disabled soldiers were looked after in a manner befitting servicemen who had risked their lives to defend the nation. Cases of court-marshalling of servicemen suspected of cowardice had attracted much concerned attention in the newspapers. Public debate called for men with shell shock to be treated away from institutional settings, particularly in vul-nerable urban centres like London. London was the first stop for the wounded returning from France, and in the summer of 1916 their very large numbers were provoking reaction from the public and the authorities.[38] Dr Haydon, writing from his Welbeck Street clinic, wanted to see the 'healthy suburbs' put to use, especially the large empty houses with 'pleasant grounds now standing idle' in Hampstead, Golders Green, and Highgate. Ordinary hospitals, he felt, would impede recovery because of the 'mental impressions from environment'. Some of these impressions may well have been sonic in nature, for Haydon also vouched for the potential 'psychological healing power of music' for men with shell shock.[39] He was a physician sensitive to sound and to favour the use of grand houses with pleasant grounds was also to favour the quiet that could be found there.

Hospitals made efforts to lift the mood of their interiors by suggesting the vitality of nature. For example, the McCaul Hospital for Officers in London experimented with a new spring colour scheme in its ward for shell-shocked soldiers. Supported by the War Office, and arguing that spring was the season of recuperation, the hospital created a sense of the outdoors in which 'the ceiling is firmament blue, while the walls are unbroken sunlight yellow. The beds and lockers are lemon yellow. The picture rail is early spring green'.[40] One feels that along with the spring colours came a restful atmosphere in these wards for soldiers.

In May 1916, this letter appeared in *The Times*:

> Sir, I have a hospital for 30 sick and wounded officers. To those suffering from shell shock quiet is most necessary. They are very weak and cannot bear much movement. A stone's-throw from the hospital is the Chelsea Physic Garden, seldom, if ever[. . .] entered by anyone except the gardeners. I asked the secretary of these gardens whether not more than four officers suffering from shell shock might have their chairs carried across and spend their afternoons there.[41]

This plea for quiet outdoor rest in the Chelsea Physic Garden next to the Thames was rejected by the garden's secretary, but it provoked over 100 letters from residents of garden squares in the west of London offering them as 'rest-places' for convalescent soldiers. Messages were received from ten squares: Belgrave, Berkeley, Cambridge, Cavendish, Grosvenor, Hanover, Kensington, Manchester, Oxford, and Portland.[42] Most of the residents of the West End squares had joined the campaign, which became known as 'Squares for the Wounded', in time for the influx of wounded from the Somme in the late summer of 1916.[43] These squares were in fact gardens and small parks with well-tended lawns, lofty trees, flower borders, and statues, surrounded by desirable addresses occupied by people of restraint. This was central London, but quite the most genteel and hushed part of the city.

Clearly it was the officer class who were deemed to be most entitled and sensitive to country-inspired therapeutic atmospheres. Much of society saw it as correct for a gentleman to be returned to his rightful place of town or country sophistication. The well-to-do and the upper-middle classes knew something of the place of quiet in healing from their family doctor, who asked for the sickroom at home to have a window facing the garden not the street.[44] Yet, such medical doctrine was also mingled with cultural assumptions about national identity. All classes had been reminded by the recruitment posters of a soldier standing tall amid the rolling beauty of southern English pasture, dotted with thatched cottages and a dovecote, that the ideal image of the nation was pastoral (Figure 2.2).[45] This country scene reflected comforting rural myths, not least that the character of the English landscape defined by restraint, reticence, and politeness was matched by the character of the English gentleman.[46] These apparent traits were not limited to officers and were appreciated and re-enacted across classes when war placed new pressures on how the nation could understand itself.

In this way, all wounded soldiers, shell-shocked or not, officers or ranks, deserved quiet: out of respect, as a return to an essential state of being, as well as for the benefits to the body and mind in recovery. These tenets were manifest

Figure 2.2 A 1915 Parliamentary Recruiting Committee poster.
(© Imperial War Museum, Art.IWM PST 11767)

in the large, embroidered letters of a banner requesting 'QUIET FOR THE WOUNDED' which was stretched across the street outside Charing Cross Hospital, London, in the first days of the war (Figure 2.3). Charing Cross was not a dedicated shell shock hospital—all recovering soldiers needed quiet. Heavy traffic was diverted away from the hospital to minimize street noise. The Defence of the Realm Act even made it an offence for Londoners to whistle at night in the time-honoured way to call a taxi in order to protect the sleep of soldiers in hospital or elsewhere from being disturbed.[47]

The model of shell shock care that members of Parliament and the wider public liked best was that shaped by Lord Knutsford, hospital reformer and philanthropist. He founded six small hospitals in and around London for officers with 'nerve exhaustion and traumatic and shock neurasthenia', to spare them the indignity of certification and being sent to an asylum.[48] While these places were called hospitals, they were essentially small stately homes that drew upon the model of the country house, even if they were in the capital. The first, lent by the late Lord Rendel's trustees, was opened in January 1915 at Palace Green, Kensington, with space for thirty-five patients accommodated in private rooms. Lord Knutsford extolled its virtues in the *Morning Post*:

Figure 2.3 A banner reading 'Quiet for the Wounded' outside Charing Cross Hospital, London, September 1914.
(© Imperial War Museum, Q53311)

The house is quiet, 'detached', overlooking Kensington Palace, with a small garden of its own. It could not be better[. . .] The Hospital is to be called 'The Special Hospital for Officers' as we are anxious not unnecessarily to emphasise to its inmates that they are suffering from shock or nervous breakdown.[49]

Knutsford made it plain that the existing hospitals in London did not have the beds nor the facilities required of officers with nervous breakdown, not least of which was 'absolute quiet and isolation in separate wards'. His mission was to create environments that would directly address this need.[50] The atmosphere of his Special Hospital for Officers was noticed and felt to be appropriate by *The Times*:

It would not be easy to find a more sequestered and restful spot in the midst of a great city. Within sight are lofty trees, green spaces and the time mellowed brick of Kensington Palace—as much tranquil old world charm, perhaps as survives anywhere in London.[51]

Two months after the Palace Green opening, demand was such that Sir Leicester Harmsworth gave his 'beautiful house, Moray Lodge, and garden on Camden Hill' for use as a similar kind of hospital. The War Office asked for more grand houses like this to be opened, resulting in the Empire Hospital in Vincent Square, Templeton House at Wimbledon and Latchmere House at Ham Common. Knutsford reported that 'Misses Alexander gave us the use of Aubrey House, and as its beautiful garden adjoined the Moray Lodge grounds nothing could have been more desirable'.[52] The twenty officers convalescing there were to be 'lured back to health of body and mind by the beauty of their surrounding' in formal gardens that were among the largest in London.[53] All of these small but emblematic centres of shell-shock care were selected for their leafy settings of calm peace and quiet.

True quiet found in the countryside

A marked anti-urban tenor to the debate about shell shock recovery was evident. Some argued that shell-shocked men should not be treated in London at all, where the threat of air raids was ever present. The chaplain of one of the large military hospitals wanted to see men moved to the 'quiet and fresh air of the country' to be able to rest their nerves.[54] Men with shell shock were especially disturbed by the possibility and occurrence of air raids and were best kept out of London in order to be able to recover properly, wrote playwright Seymour Hicks from the smoky calm of the Garrick Club.[55] When the Duke of Sutherland launched an appeal for country houses suitable to be used as hospitals or convalescent homes, he was overwhelmed with offers. Almost immediately 250 country houses, manors, and halls were put forward, though few were found to be suitable.[56]

The idea that connection to rural life would be beneficial was given further authority in 1917. Thomas Lumsden, a doctor and medical referee for pensions, developed a scheme for discharged, pensioned men who were still suffering mental ill health. Lumsden wanted these men to be 'sent to live at a country gentleman's house, or with a farmer or gardener, where he would be well fed, and be encouraged to perform a small but increasing amount of simple work, under absolutely safe conditions, in the open air, and with a peaceful, natural environment'.[57] A week after promoting his plan in *The Times* he was inundated with offers of hospitality for pensioned servicemen and his project called the Country Host Scheme took off.[58] A conference of neurologists decided at the War Office that only severe cases of shell shock in men who were not returning to the fight should be eligible.[59] There may well have been concerns that retreat to the countryside would create conditions that were not conducive to a return to duty. After the war in 1920 Lumsden criticized treatment centres like Maghull and Seale Hayne for failing to permanently cure shell shock patients with 'psycho-therapeutical treatment' alone. Under his scheme, which was now run by the Red Cross, all cases must be 'sent to the country' for at least three months to benefit from the peaceful environment.[60]

The requisitioning of health spa institutions around Britain provides further evidence of the status of outdoor quietude for psychological recovery.[61] These Victorian institutions were founded for their healthy position away from urban centres. Craiglockhart hospital just outside Edinburgh was one such establishment made famous for its poet patients, Siegfried Sassoon and Wilfred Owen, and W. H. R. River's psychotherapeutic techniques, featured in Pat Barker's novels. As an earlier hydropathic hotel for Edinburgh's worried wealthy, it offered 'its residents all the amenities and retirement of quiet country life'. There were gardens and grounds extending to twelve acres.[62] In wartime, officers voiced their appreciation of the setting in their stories, drawings, and poems that were published in the home-made hospital magazine *The Hydra*. James Butlin, a Lieutenant in the Yorkshire and Dorsetshire Regiments, found however that ' "a complete and glorious loaf" palls after a few days'. He was soon bored 'doing a little gardening and poultry farming after breakfast' with 'fretwork and photography after lunch' followed by 'viewing natural scenery after tea'.[63] Perhaps the medicine was working, but during wartime it felt incongruous to be spending one's time in this way. Another patient, who signed himself as North British in *The Hydra*, felt the benefits of his Craiglockhart treatment, offering these verses to his colleagues:

Be still, my soul, and sound
The stillness all around:
Brings peace to wearied souls,
'Tis Nature's Paradise.
Be still, my soul, and rest,
The peace that was your quest
Is here, in this demesne:
'Tis Nature's Paradise.
Be still, my soul, and sleep,
Refreshing, good, and deep:
And waking you will say
'Tis Nature's Paradise.
Be still, my soul, delight
In all the joys that might
Be yours, whilst you remain
In Nature's Paradise.[64]

This poem makes plain that nature was a true source of quiet, rest, sleep, and delight for one soldier at least. The turbulence of the mind could find stillness. A spa hospital like Craiglockhart could offer this in ways that conventional hospitals and asylums definitively could not. But, the opportunity to feel this way was largely an officer's privilege. Lord Knutsford's special hospitals by favouring officers reinforced existing class inequalities in relation to diagnosis and treatment.[65] The Knutsford care model could never be scaled up and transposed across the social divide, but the lore of these hospitals informed the demands of advocates of mentally disabled ex-servicemen over the next few years, who proposed similar small-scale therapeutic communities for the soldiering masses.[66] A village community, not a hospital community, was the one that inspired the medical and political authorities the most.

The 'beneficent alluring quietude' of the Village Centre utopia

The Enham Village Centre in the Hampshire countryside was a particularly interesting response to the need to rehabilitate shell-shocked men. Historians have not yet explored the phenomenon of post-war Village Centres. A sister project called the Papworth Village Settlement in Cambridgeshire, established in 1917 for families with tuberculosis to live and work in was configured as 'a return to Nature'.[67] The vision for Village Centres was formulated towards the end of 1917 when the treatment of shell shock was a continuing national concern and there was still much uncertainty about the proper state provision for disabled soldiers. A driving motive for the establishment of Enham Village Centre was the desire by men like the Conservative politician and lunacy reformer Sir Frederick Milner to prevent soldiers with shell shock, especially rank-and-file men, from being incarcerated in asylums.[68] He campaigned widely around Britain to raise funds for Enham and other organizations including the Ex-Services Welfare Society.[69]

Enham offered help to all disabled servicemen but it was understood that neurasthenic cases would form a large part of the intake.[70] This was to be a place where medical rehabilitation was on offer, combined with gentle outdoor or craft work, and a distinctly non-institutional environment.[71] Warwick Draper, speaking for the Village Centres Committee, spelt out the kind of holistic atmosphere that would be required: 'cheerful and tranquil out-of-door surroundings' and a 'quiet healthy environment'.[72] This kind of countrified sonic milieu would facilitate a treatment routine that would focus essentially on the well-known rest cure interventions of 'wholesome food', with 'massage, electricity and baths'. On top of this, gentle 'graded work' would have its own rehabilitative effects so long as it was not of an industrial kind that reflected the 'speeding up' of the factory system.[73] A regime of pastoral work rhythms and quietude would help neurasthenic men regain their capacity for work, while they took on new skills to become independent and productive citizens.

Careful attention was paid to the selection of a special health-giving location for Enham Village Centre. Flat or damp country would not work. The best results would come from surroundings of 'natural beauty and tranquillity'.[74] These characteristics are inseparable; natural beauty seems unlikely to be perceived without an attendant peacefulness. Both are associated by Warwick Draper with health and healing. The Centre boasted 1,000 acres of countryside, including 150 acres of woodland.[75] Its promotional material, that sought to attract financial support, depicted in an aerial photograph the grand Enham Hall and its cottages in splendid rural isolation, and a flock of Hampshire Down sheep had its own page.[76] The *British Medical Journal* featured the May 1919 opening by the Minister of Pensions for fifty neurasthenic patients, noting that the estate was 'situated in a healthy part of Hampshire, with three hamlets upon it, five farms, a post office, village hall, cottages and smithy'.[77] Such imagery conjured up not a pre-war but a pre-industrial England, where time moved slowly. It called to mind a time when the national soundscape was pure and simple, and paced by the human not the machine. Stanley Baldwin employed these themes in 1924 when he spoke about England being the country and the country being England. This England came to him through his senses he told the Royal Society of St George: 'the tinkle of the hammer on the anvil in the country smithy, the corncrake on a dewy morning, the sound of the scythe against the whetstone'.[78] Baldwin went further to associate these sensations with 'the very depths of our nature', going back to 'the beginning of time and the human race'.[79] Fundamental to being English was to be steeped in rural traditions, or at least to recognize their power though few experienced such things in the towns and cities where most lived. Enham emerged from similarly hazy ideas of the authentic nation being a rural one of simple honest labour, all the more appealing in wartime.

Rural work rhythms

Surrounding the planning of Enham in 1917 and 1918 and then in its early years of activity until at least 1922 was a conviction that this new community was a model for how men should live in harmony with the natural world and make their livelihood there. From some intellectual quarters there were humane strands to this idea, tinged with pastoral utopianism. There were pragmatic medical voices who saw a need for damaged men to work apart from the intensities of industrial life. And, perhaps surprisingly, there was senior military opinion that a long-term solution to the mental incapacity of shell-shocked soldiers would have to include opportunities for these men to make their living outdoors in the countryside. These three strands of thinking all shared a belief in the curative value of craft-work with its gently productive rhythms that stood apart from gruelling factory regimes. In the face of a decimated post-war economy and hundreds of thousands of demobilized ex-servicemen in search of any work they could find, this proposition appears somewhat quaint.[80] However, while shell-shocked and disabled soldiers were not exempt from these personal and national concerns, those closest to their rehabilitation considered they deserved and needed special consideration, at least in the immediate aftermath of the Armistice. After all, in relation to the population of working men, this relatively small cohort were not expected to be an engine of the economy. Yet, how they were treated reflected a wider interest in how all people affected by the war would in the ideal world have been given sufficient time and space to recover their strength.

The government-sponsored quarterly magazine, *Reveille*, billed as the first journal devoted to the 'care, re-education, and return to civil life of disabled soldiers and sailors', put the case for the shell-shocked man's need for nature.[81] Under the editorship of the novelist and playwright John Galsworthy it incorporated literary contributions from soldiers and Galsworthy's contemporaries, together with articles from the War Office and Ministry of Pensions, and pieces about new orthopaedic centres and living with deafness. This curious compote was intended to champion the restoration of the 'spirit no less than the body' of disabled men and their carers by connecting them to countryside.[82] Enham is featured in *Reveille*. Galsworthy gave his sentimentalism free rein in the cause of the Village Centre. In this typical piece of his own writing in *Reveille*, he attempts to argue that severely injured soldiers need the sensory tonic of the natural world to find contentment again:

> Every tree is dowered with young beauty, and no two the same. Last evening they stood against the sunset, magical and delicate, with pale gold light between the curling quiet leaves; far away on the skyline some elms had their topsails set; the birds had lost their senses, singing. In such moments this green Land of ours has incomparable beauty, seeming to promise happiness which can satisfy even the human heart. The thoughts of wounds, of disfigurement, and blindness, of lost limbs, twisted limbs, the thousand and one bodily disasters which this war has brought, becomes unbearable unless we keep the hope and the will to give back to the wounded something of this Spring and of the Summer which Spring leads to.[83]

There is redemption for the injured and for the nation in the richness of burgeoning nature, Galsworthy argues. Nature's impressions get inside us and nourish the mind: 'And life should have its covering of dream—bird's flight, bird's song, wind in the ash trees and the corn, tall lilies glistening, the evening shadows slanting out, the night murmuring of waters.'[84] Galsworthy publishes an extraordinary photograph of a beaming convalescent soldier, charming and charmed by white doves (Figure 2.4). The nameless 'Happy Warrior' is in the courtyard of the Old Heritage Craft Schools, Chailey, Sussex, and captioned with 'Laetus sorte meâ' (Happy in my lot). The message is that this damaged man—we know his left arm is wounded, his mind may be too—was being cured through this intimate communication with the soft caresses of a pre-industrial England.

Galsworthy rejected the prevailing urban and industrial pattern of British life which alienated so many: 'the towns have got us—nearly all', he wrote in his magazine. 'Not until we let beauty and the quiet voice of the fields, and the scent of clover creep again into our nerves, shall we begin to build Jerusalem and learn peacefulness once more.'[85] Submitting to nature's soft sensory stimuli was a way for all Britons to sustain themselves in modern times.

In 1917, the Village Centre planning committee, which comprised a group of esteemed medical men including the psychiatrist George Savage and the surgeon Wilfred Trotter, wanted Enham to be the first of a series of 'settlements' that would combine 'restorative treatment with industrial and social reconstruction'. These committee members were more pragmatic than Galsworthy, but they too foresaw scope for social reforms based around village communities. It was not a return to feudal ways that was wanted, an editorial in *The Lancet* pointed out, but a recognition of 'the widespread value of happiness' that can come from the 'practical business of agriculture'.[86] They wanted to see men with psychological damage 'trained for rural life and village crafts, believing that in a congenial rural

Figure 2.4 The 'Happy Warrior' in recovery at Chailey Craft Schools, Sussex.
(From *Reveille*, 1 August 1918)

life lies the cure for their physical and mental ills'.[87] The industrial workplace was ill-fitting for these men and, it is implied, for many who wanted to contribute to modern society but not within the confines of the prescribed workplace systems.

> Factories will doubtless remain in certain industries. But there is a growing reaction against the universal recourse to the factory system, with its specialisation of labour, speeding-up, and lack of personal relationships. There is in many quarters a desire to return to a modified form of the old craft guilds, in which a man may learn to create articles for use of a high standard and himself become

a master craftsman, not wholly spending his life in performing some mechanical and deadening process.[88]

In 1920, the craft-work at Enham involved the rhythms and repetitions of carpentry, basket making, upholstery, and electrical fitting. The horticulture and farm work were light: planting seedlings, hedge clipping, and fruit picking.[89] Inherent in this work were smooth, predictable movements and sensations. Doctors worried during the period of the war not only about shell shock but more prosaically about the effect of city din, especially of traffic and motor horns, which could damage the nerves and drain energy because of their lack of rhythm.[90] Civilization's machinery was pathologized not simply for its noise but because of its startling arrhythmias. But it was industrial working conditions that worried doctors most, including those who had been planning Enham before it opened. In 1913 the Home Office had set up a survey into the problem of industrial fatigue and 1914 saw the creation of the Health and Munition Workers Committee with a remit to investigate the health and efficiency of workers using physiological and psychological methods.[91] This scientific analysis of work practices intensified throughout and after the war, taking an interest in the effects of work pace and rhythm, and the environmental factors of temperature, humidity, noise, and light on the welfare and productivity of human bodies.[92] It is likely that the therapeutic programme at Enham grew from this thinking, thinking that took seriously the influence of sensory stimulation on well-being. At Enham, there was relative quiet. The possibility of sensory overload was minimized and a world away from the dust, heat, and roar of Britain's staple industries (textiles, coal, iron and steel, and heavy engineering). The natural rhythms of the working body were accommodated, not overstretched.

Anti-industrial sentiment after the war was voiced too by senior military figures. Field Marshal Douglas Haig, the commander of the costly battles at the Somme, Passchendaele and the concluding Hundred Days Offensive, took an active interest in Enham's work. Haig emphasized the need for a permanent rural community at Enham because 'there are a large number of men so badly injured by war service that they will never be able to return to the ranks of the industrial world'.[93] Major C. Reginald Harding believed that work on the land was right for men 'who suffer from nervous and mental shock and exhaustion'. He argued that 'these above all need the restful and healing influence of quiet places and occupations in the open air and mental treatment of the right kind'.[94] Harding and Haig believed the Village Centre idea would have a permanent place in Britain because it chimed with the revival of village and rural life that was in the air in the aftermath of war. 'We know that the war has turned men's minds towards a country life', wrote Harding.[95]

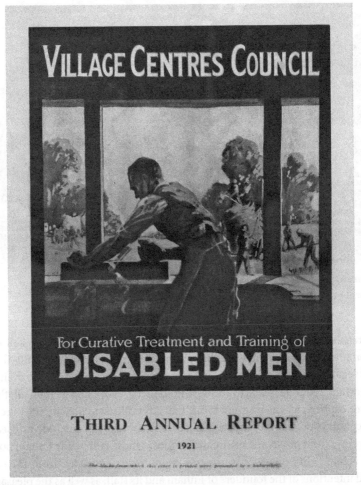

Figure 2.5 The 1921 annual report for Enham Village Centre.
(Enham Trust archive)

In Enham Village Centre, then, we see faith placed in the peace and quiet of the countryside as a reliable and tested way to rebalance minds. The traditional rhythms of making things out of wicker and wood, picking fruit, and cultivating the land without machinery were believed to bring veterans back into a natural synchrony with age-old practices, while providing a simple soundscape of known and reassuring clicks and swishes (Figures 2.5 and 2.6).[96] We have a limited vocabulary for such small sounds of human interaction with nature and with natural materials, but I suggest that they are part of a mood of quietude that appeared to have particular therapeutic purpose after the violations

Figure 2.6 Basket making at Enham Village Centre.
(From the 1923 Village Centres Council Annual Report, Enham Trust archive)

of trench experience. Visitors to Enham were sensitive to the charms of its environment. Though there were thirty acres of gardens devoted to fruit, flowers, and vegetables, *Fruit Grower* magazine managed to locate something even more important about the place to recovering soldiers: 'above and beyond all, there is an atmosphere, an elusive something, that has nothing to do with the pure air of the countryside, but is composed of a beneficent alluring quietude infused into the centre from the serene spirituality of those whose faith made Enham possible and actual'.[97] In other words, the healing quietude emanated from the careful good intentions of the founders of Enham and its staff, as well as the pleasant rurality of its geography.

By 1922, after four years of work, 725 patients had been treated at Enham. At that time there were 103 men resident at the centre, forty-six 'were suffering from neurasthenia' and thirty-five from surgical conditions, many of which were gunshot wounds to the head.[98] The remainder had presumably improved sufficiently to be discharged. However, the vision of building a healthy and prosperous post-war rural society with the guardianship of nature's calming balm did indeed transpire. The 1935 annual report underlined the continued focus on establishing a permanent settlement. By then some 400 men and their families were living and working in the village, in sixty-two 'memorial cottages'. Cottage building was funded by the Prince of Wales, Princess Mary, and the British Legion. Selling basketry, furniture, and farm produce had become a commercial operation.[99]

Men still struggled with the effects of the war in the 1930s. Doug Saunders was born at Enham in 1928, the last child of six, and he told me about his father when I visited him in the village in 2017. Doug's father was a professional soldier and bandsman who was offered rehabilitation at Enham by a medical assessor. He moved there with his wife and three children in 1919 and became a cobbler. Doug remembered his father could be frightened to go upstairs and sometimes needed to disappear to the bottom of the garden. Doug reported seeing men have fits at Enham, one of whom had a metal plate in his head. Another man marched up and down in the upholstery workshop and two killed themselves on a nearby railway line in the 1930s.

The pastoral approach to shell shock treatment had been tried out before Enham at Seale-Hayne neurological hospital in Devon. The physician in charge, Major Arthur Hurst, took on rank-and-file men, some of whom were previously deemed incurable, with the intention of 'rapid cure' and return to the Front.[100] This was not a community like Enham but a system of persuasive treatment that culminated after a few weeks in a period of 'open-air employment', which included farm work, gardening, and handicrafts in 'beautiful surroundings'.[101] For Hurst, quiet was important and the absence of air raids at Seale-Hayne was an essential constituent.[102] The last phase of his regime to get men back to the fight involved outdoor work 'on the farm'. Uniquely, Hurst documented the results of his work on film in 1917 and 1918.[103] We see Private King and Private Sandall leading a bull through a meadow and digging a field. Private Read's swaying hysterical gait and nose-wiping tic have vanished and now he is shown feeding chickens. There is a group picking blackberries and two men at work in the fields with horse-drawn ploughs.[104] If this was a film intended to advertise the success of his curative protocol it also demonstrated the belief that immersion, though very briefly, in a quiet pastoral work routines would be the final stage in curing shell-shocked men. How long men survived on active service having exercised their blackberrying abilities is not known, though Hurst claimed this kind of treatment reduced the likelihood of relapse.[105] The idea that pastoral quiet would be efficacious at all during wartime is remarkable in itself. The scheme of treatment at Enham was more considered than at Seale-Hayne and geared towards developing a self-sustaining community who would live on after the traumas of war.

Quiet for national recovery

Most men and women who had served did not have the chance to enjoy the quietude of the country. They had to make the best of what was around them. Historian Denis Winter has argued that most who returned home to the towns and cities required 'a period of quietness, a second adolescence as it were, to shed the past and get back into life at a lower key than they were used to'.[106] It was a matter of going back a step or two to be able to go forward again. Some withdrew into solitude or nomadism, while others resorted to angry demands from society for recompense. There were few jobs about, but in any case a return to former employment might be delayed because men needed time to find their feet and were happy for their country to look after them for a while.[107] Yet in the difficult period of readjustment, which could take many years, simple, relaxing routines were valued. Denis Winter's father had fought, and he wrote in his unpublished memoir: 'I was on the dole right through the summer, revelling in the open-air swimming pool of Victoria Park, reading in Hackney Library, rowing on the Lea, assiduously practising the violin. The springs were unwinding.'[108] There was, it was said, a new kind of hushed atmosphere in the years after the war was over. In October 1919, Robert Graves found Oxford University 'remarkably quiet'. The expected boisterous behaviour of the old days was muted in both returning soldiers and in new boys coming straight from public school.[109]

We know from the first chapter that the war changed soldier's listening practices and that the meanings of sounds had been altered by that experience. Loud noises, the sound of birdsong and gaps in sound—the silences—could all be read differently now. Back in Britain, even after the war had ended, these changes reached into the private and public life of many civilian veterans, their families and others who had remained at home. New sensibilities were brought home, though how persistent these changes may have been in peacetime, in civilian life, is hard to say. It is clear, however, that a community far larger than the returning fighting forces had been engaged in new ways of listening for danger, its passing and presumably its opposites. Civilians in the southeast of England had found the insistent throbbing of the guns across the Channel, 'like the thud of some giant propeller', invaded walks in the countryside, penetrated inside homes and reminded families of their loved ones under fire.[110] For Sassoon on leave at his family home in the summer of 1917 there was no escape: 'I'm going stark, staring mad because of the guns', he wrote.[111] Despite the efforts of the poet Mary Beazley to get away from the sound, she still thought she could hear the guns 'in the plash of a wave' and 'the sea-birds' call'.[112]

Civilians in many urban areas of Britain and in towns along the south and east coasts of England had cultivated ways of 'listening out' for airship raids from late 1914. For their own safety and as a civic obligation, civilians were asked by authorities to listen out for all kinds of new warning sounds, from 'buzzers' to 'hooters' and 'syrens'.[113] However, in London, by-and-large, little sonic warning was given, especially at night-time. It was left to Londoners to listen in the dark for themselves and by 1917 bombing raids by Gotha and Giant aircraft as well as Zeppelin airships were proving fatal and causing much fear.[114] 'The Gotha hum'—'quite unlike any other machine noise'—was a sound strongly associated with trepidation.[115] The rumbling and roaring of these raids, whether German aircraft and bombs or British guns, were still remembered twenty years later.[116] They had travelled deep into the mind and changed listening and its connotations. The sonic skills and sensitivities that had been developed in the trenches were perhaps not much use in civilian society, but they would persist. Robert Graves realized this in 1919 when he found himself flat on his face, taking cover automatically from the sound of a car backfiring.[117]

One way to find some peace and rebalance the mind was to create a small place of one's own, away from urban life. Assemblages of higgledy-piggledy 'plotland' homes appeared in the south-east of Britain after the war, spurred by the chance to be independent, without the need for building societies or even builders when jobs and money were so tight for many.[118] A simple plotland home in the countryside provided a kind of home-made rest cure on the land for ex-servicemen and their families. Dreams of chicken farming or market gardening may have been easily shattered, 'but the patch of land and the owner-built house on it remained as some kind of security' and comfort.[119] Attempting to strip away the nostalgia of memory, Dennis Hardy and Colin Ward have found that 'the recollection of plotland people is a simple tale of quiet enjoyment' and healthy living centred around the fresh air, local farm produce, and the 'tranquillity of a riverside haven or woodland setting'.[120] This makeshift exploration of the pastoral was available to anyone with ingenuity and imagination.

Socially conscious medical men raised the importance of quiet rest for overwrought middle-class professionals after the war. Edwin Ash in 1920 voiced concern that the entire working nation was at risk of nervous breakdown from the pressures of modern life. Previously it was cultivated writers, artists, and scholars who were thought to have the most susceptible nerves. Now urban 'brain-workers'—lawyers, social reformers, vicars, journalists, physicians— were said by Ash to be particularly vulnerable to nervous instability and this threatened the national economy. Ash was a nerve specialist and he recommended that state-funded rural sanatoria were set up to help these kinds of people to regain their strength for the advantage of everyone.[121] The Scottish surgeon and ear, nose, and throat specialist, Dan McKenzie, also warned of the dangers of city noise while explaining the important kind of quiet offered by the natural world and the stirring sounds of the wind in the trees or the waves on the cliffs. Such sounds had a deep 'inner meaning' for humans, he argued; they were 'ineffable'.[122] In his book *Aromatics of the Soul* (1923) McKenzie contrasted the 'breath of hot oil and metal' of the modern world with the sounds and smells of the farm which were able to relax the body and allow the mind to drift happily:[123]

How warm and comforting is the smell of a byre full of cows! Plunge into it from the cool of the evening and listen again to the sudden swish of the warm milk into the pail, the uncompleted low of the sober cattle and the rattle of the chain as they turn to look at the new-comer. A gentle relaxation of the spirit attends the visit like the relief of the limbs from a cramped position, and we readily fall into that mood, so rare these latter days, when attention disperses and the reins drop on the neck of the mind so that it wanders on at its will up and down the lanes and by-ways of fancy.[124]

Few had access to the sensory world of the farm, but McKenzie's imagery and his rhapsody about the soundscape of the rural calls to attention the place of English pastoral music in British culture at this time. Composers such as Edward Elgar, Arthur Bliss, George Butterworth, Frederick Delius, and perhaps most notably Ralph Vaughan Williams were making music that quite intentionally drew upon several key cultural inspirations referencing the pastoral. These were: folk music tunes and lyrics; poetry with rustic imagery, by Thomas Hardy and A. E. Housman, for example; and the evocation of the sounds of nature in onomatopoeic reference to water, wind, or bird calls.[125] The moods, work cultures, folk songs, landscapes, and acoustics of the countryside could be expressed in orchestral music—the feel of nature could be articulated through this form of music. For instance, the opening of the first movement of Vaughan Williams's Fifth Symphony with its soft horns seems to 'summon up an archetypal English landscape of summer pastures and distant hills', David Matthews has argued.[126] In The Lark Ascending, 'the dialogue between the solitary listener and nature—the lark—becomes, at the end, a monologue for the lark, the solo violin, who leaves the orchestra behind to climb up alone in the clear air'. The lark's song has been rendered by a human musician, Matthews points out, and in this way Vaughan Williams tells of the interdependence of man and the natural world.

While some of the most prominent British critics of the period argued passionately against composers evoking conspicuous markers of provincial 'Englishness', which after the awful events of WWI could be meaningless vestiges of Romantic culture, pastoral music found appeal and success among many listeners, Eric Saylor has argued.[127] In the 1920s and 30s concerts are frequently found in the listings of local and regional newspapers and the BBC plays pastoral music regularly as part of its classical programme. Composers of pastoral works were concerned with the creation of a national sound rooted in the English landscape, its creatures, and its folk song. But this music, though it could provide moments of escapism and nostalgia, was also able to reckon with the pain and death of world war, making afresh, as Eric Saylor has contended, 'traditional literary and musical tropes of pastoralism in profoundly modern ways.'[128] Pastoral music encouraged not necessarily an *escape* from the complexities of modernity but rather an *exploration* of it in contemporary cultural contexts.[129]

In the years before the war, there was political and non-political unity around the idea that the land, smallholding and agricultural work, and the countryside itself, were 'coming to represent order, stability and naturalness'.[130] Nature's everyday sounds and those of rural work implied a modern rationality of order and health when compared to the ravages of Victorian industry. After the war and the challenge to civilization, the most essential parts of British culture needed to be identified, retained and transmitted to reconstruct the nation.[131] The peaceful countryside was definitively part of this culture, even if this was a rhetoric expressed chiefly by urbanites and led most vocally in the post-war years by Stanley Baldwin's aesthetic ideals of rural Englishness.[132] Beatrix Potter's animal tales and Kenneth Grahame's *Wind in the Willows* were bedrocks of this culture, present throughout the war and needed all the more afterwards. Ernest Pulbrook epitomized in *The English Countryside* (1915) a mawkishly romanticized vision of rural life, telling of the peace and beauty of the landscape and the lessons of patience and hope it taught. His vision of rural England is built on 'ripeness and repose', 'ordered peace and settlement', 'tranquil spots where Nature holds sway', 'the soughing of the pine woods', 'the quavering bleat of new lambs', 'the song of the stream', and so on.[133] Before and during the war, the quietude of rural England was firmly part of political and literary culture that was drawn ever closer after war. These comforting sonic traditions and familiarities of pre-war life relied on myth, but they were required for a society trying to find meaning in what had happened and recover.[134]

In November 1919, a short abolition of all sound was chosen to mark the end of the war the year before and to allow those still living to commemorate and remember who had been lost. The two minutes of Armistice silence was a most unnatural pause, its strangeness coming from the simultaneous curtailment of all sound and movement in Britain and throughout the Empire.[135] This was the sound of the realm without humans, 'devoid of all the banality of speech' apart from the cooing of Town Hall pigeons.[136] On 12 November 1919 the *Manchester Guardian* reported the previous day's silence:

> The first stroke of eleven produced a magical effect. The tram cars glided into stillness, motors ceased to cough and fume and stopped dead, and the mighty-limbed dray-horses hunched back upon their loads and stopped also, seeming to do it of their own volition . . . Someone took off his hat, and with nervous hesitancy the rest of the men bowed their heads also. Here and there an old soldier could be detected slipping unconsciously into the posture of 'attention'. An elderly woman, not far away, wiped her eyes and the man beside her looked white and stern. Everyone stood very still . . . The hush deepened. It had spread over

the whole city and become so pronounced as to impress one with a sense of audibility. It was . . . a silence which was almost pain . . . And the spirit of memory brooded over it all.[137]

The silence was, of course, too much to bear. It could not be maintained. As *The Times* reported in 1920, the silence was broken by sobs here and there. In the deepening quiet a woman's shriek 'rose and fell and rose again' until the silence 'bore down once more'. After a few years, the silence was being interrupted by the sound of trucks and workplace sirens.[138]

Nevertheless, the status and meanings of silence in civic society altered after this event, though to pinpoint the character of this change is difficult.[139] One indication of the continuing significance of this liturgical silence was its broadcast to the nation from the outset of John Reith's BBC. In November 1923, it was accompanied by 'The Last Post' and 'Reveille'. The broadcast was extended to include 'O God, our Help in Ages Past' in 1924 and the Lord's Prayer the following year; also in 1925, the sounds of the crowd gathering in Trafalgar Square were heard over the airwaves for two minutes beforehand, probably to heighten the sense of contrast with the stillness that followed.[140] As Adrian Gregory observes, the easiest option for the BBC would have been simply to stop broadcasting for the requisite two minutes. However, Reith's team was concerned not only to transport the listener to the vicinity of the Cenotaph on Whitehall in London, but also to convey a sense of both sublimity and communitas. The extraordinary effort and technological investment necessary to coordinate the silence across the regions by the strokes of Big Ben shows how important the BBC believed a concurrent emotional experience of the silence would be, and that it could orient listener's minds at this moment towards the political and constitutional heart of the nation.[141]

What was heard was the sound of the world when humanity had withdrawn. Men and women and children heard the natural state of the world as they froze for those two minutes. They were still present, but only as listening ears. And at home on the wireless, listeners there heard the whoosh of static and perhaps the waft of wind across the microphone. An account published in the *Radio Times* in 1935 revealed more of what the silence at the Cenotaph contained:

> the silence is not a dead silence, for Big Ben strikes the hour, and then the bickering of sparrows, the crisp rustle of falling leaves, the creasing of pigeons' wings as they take flight, uneasy at the strange hush, contrast with the traffic din of London some minutes before. Naturally vigilant control of the microphone is essential. The muffled sobs of distressed onlookers, for instance. Audible distress too near a microphone would create a picture out of perspective as regards the crowd's solemn impassivity and feelings.

> Our job is to reduce all local noises to the right proportions, so that the
> Silence may be heard for what it really is, a solvent which destroys personality
> and gives us leave to be great and universal.[142]

Those words at the end of this quotation echoed John Reith's ideas of what broadcast radio should achieve with its cultural paternalism, something that will be
explored in the next chapter. However, for now I want to note how the BBC's
microphone had allowed all sorts of natural sounds to creep into the silence.
Sparrows, leaves, and the bustle of pigeons are here granted permission to inhabit even the most sacred of silences, making it palatable. Quiet and silence are
so closely related in human perception to the sounds of the natural world that
sometimes they become interchangeable. Quiet is nature's sound and nature's
sound is quiet, to play with Baldwin's injunction. More than this, though, in the
years and even decades after WWI, all these closely inter-related sounds of nature are inextricably tied to notions of healing and recovery.

Conclusion

The quietude associated with the English countryside was brought to the fore of
British culture during WWI and its aftermath. Those who could find it benefited
from a spell in nature's peacefulness. It was a milieu that made sense when recovering from the violence of war and it reflected the spirit of nervous pacifism in
the air during the years after the conflict. The authorities, especially in retrospect,
saw sense in drawing on the qualities and traditions of the English countryside to
heal the psychologically wounded. There was little else to turn to; the quiet pastoral provided a medical and political refuge from the bewildering crisis of shell
shock. To military, medical, and government officials, quite recovery in reach of
the natural world was a rational response to the chaos and damage wrought by
industrial warfare. It was tenable because of deeply held beliefs in the healing
effects of the nation's land and green places alongside medical reliance on the
Victorian tradition of rest cure for nervous illness including neurasthenia. The
'peaceable place on the other side' of Samuel Hynes's 'great chasm' was a pre-war
haven that society reached back to. It contained ideas far too precious to consign
to history.[143] It was necessary to step back to be able go forward again.

During the war, public sympathy for shell-shocked officers was evidenced by
a wave of offers for them to enjoy leafy London squares and country houses away
from the metropolis. While country house therapy for officers was a mark of class
privilege, it also reflected beliefs about the moral purity of the countryside as the
locus of healing. The tranquillity of healing environments was key. After the war,
there was a noticeable anti-urban and anti-industrial tone to debates about how

Britain should care for its ex-servicemen and recover more generally as a nation. Working gently on the land and craft work were idealized as ways to rejuvenate and return to function disabled bodies and minds, in a post-war world in which industrial civilization had so obviously failed. Many held a vision of society in which men worked quietly and productively outside the industrial system. This was not a romantic or nostalgic tendency so much as a pragmatic solution to sustain a working nation coming to terms with how best to progress. The rhythms of work, and the sounds of work, in the newly conceived therapeutic community of Enham Village Centre were of a certain timbre designed to be different to those of the city or factory. Enham's idyllic picture of the peaceful nation in recovery is particularly striking and long-lived. While this was the only Village Centre to be established, it thrived and grew during the interwar period as its shell-shocked rank-and-file men had families, some of whom still live there today.

In these environments of recovery from the shock of war, the qualities and textures of quiet can be detected in the reasoning, practices and locations, even though the sonic is not always explicit. Peace and quiet pervades newspaper coverage, photography, and medical commentary. In this analysis, then, I have brought to attention the assumed and implicit sounds prescribed for recovery. In doing so, the kind of quietude that soldiers and ex-soldiers are placed within is revealed to be that which emanates from the natural world: from gardens, parks, stately home grounds, plots of farmland, thatched cottage environs, and pre-eminently 'the countryside'. In the post-war period it is evident that quiet continued to be valued, needed not just for healing damaged minds but to re-civilize a nation anxious to comprehend and leave behind the brutality it had participated in. Quiet was canonized by the Armistice silence, but it was sought out every day in private moments. After all, Britain had a new sonic awareness after the war. Several million men and women brought back with them new sensitivities to sound as they returned to civilian life. Those who had spent the war on the home front in the southeast of England had also become far more sonic minded as for four years they had listened out for air raids and heard the artillery rumbling across the Channel. In the next chapter, the sonic aftershocks of WWI will be considered in the establishment of Britain's first broadcasting efforts, as will the place of nature's sounds on air in that new media context.

3

Broadcasting nature

Five years hence, the experts say, many of us, instead of motoring,
will be flying. Everything in England is to move forward except the
Church as by law established.
　　J. W. Robertson Scott, *England's Green and Pleasant Land* (1925)[1]

BBC broadcasting in 1922 brought forth a new national listening service. This
was the first electronic mass medium in Britain, and it was taken up with relish.
By the end of 1924, there were already over 2 million people listening to the wireless in homes around the country.[2] The general manager of the BBC, John Reith,
established a concept of 'public service broadcasting', which meant that public
opinion mattered, however paternalistic initial programming was, but also that
listeners' thoughts and opinions were malleable and therefore susceptible to
the power of radio.[3] Reith's aim was to bring the best of culture to all corners of
Britain and this culture included British nature.

This chapter explores how conceptions of the new domestic technology of
radio were intertwined with beliefs about the peacefulness and harmony of the
natural world in the first years of BBC broadcasting from 1922 to 1927. Reith's
vision for the BBC as a public service made use of the sonic energies of nature
to convey the full potential of what broadcasting could offer the nation. This potential was demonstrated by the transmission of birdsong, in the revelation of
nature's unheard voices, and in the chance to connect public minds to the order,
stillness and harmony of the cosmos. Could the sounds of the everyday British
microcosm alongside the heavenly arrangements of the macrocosm sooth post-
war minds, and help establish the new medium as a force for cultural enlightenment? And how could silence, of all sounds, be part of a new broadcasting
medium?

I argue that the well-known nightingale broadcast served to define a small but
essential part of Reith's vision for his public service broadcasting. Reith believed
that the public response to the nightingale demonstrated how broadcasting
could do more than convey human culture, that it could be transcendent, even
suggestive of a sublime silence for reflection that all Britons were in need of from
time to time. Moreover, many cultural commentators found within the ideas of

Listening to British Nature. Michael Guida, Oxford University Press. © Oxford University Press 2022.
DOI: 10.1093/oso/9780190085537.003.0004

the physics of radio communication a kind of cosmic nature-theology, in which listening to the radio put all souls in all nations in direct and instantaneous contact with the heavens. The postwar social context is significant in the shaping of such views. The media historian David Hendy points out that the early BBC was shaped by personal wartime struggles that mean it can be thought of as an institution shaped by 'systems of feeling', as much as by rational policy-making.[4] So, while the views of broadcasting experts and critics together with the listening public are the main voices to be marshalled here, the moods and emotions of the post-war period are in attendance.

John Reith's public service nightingale

Voices! Voices! The voices of a mighty multitude, year in, year out, holyday and holiday, noon and night, flow over our heads and under our feet in a ceaseless, silent chorus.[5]

New electric media have always been imagined to be full of unregulated conversation. In 'Telephone London' the journalist Henry Thompson worried about the endless telegraph and telephone messages fluxing across cities and nations. In 1901, voices were contained in the privacy of the cabling, but radio and then broadcasting brought yet further concerns of communication overload as voices became airborne. Even so, John Reith was able to think of his BBC broadcasting as 'the voice from the silence'.[6] The BBC voice was clearly not just another from the cacophony. It was considered to be the first true voice to emerge from the ether. Reith's BBC voice was singular and authoritative, and distinct from the chatter of amateur wireless fanatics and the stream of cheap newspaper stories. He argued that his broadcasters, with their own personalities and intimate engagements with listeners via the microphone, were the essence of this new carefully considered voice. To speak to the nation was an honourable role 'that only those of exceptional and proved capacity are fit to aspire to', Reith wrote.[7] The broadcaster was a 'guide, philosopher and friend', but above all someone of imagination.[8] The imaginative character of broadcasting was put into action from the beginning of the BBC to bring a unique voice to the airwaves.

In May 1924, a still-iconic and fondly recalled broadcast event took place, when a nightingale was heard to sing along with Beatrice Harrison's cello performance in her Surrey garden. Scholars have tended to account for this episode as a media stunt, technical achievement, or historical curiosity rather than assigning it any more significant cultural meaning in relation to the new project of public service broadcasting.[9] To understand the nightingale broadcast requires immersion in the minds of the men who established the service and strove to gain public and political acceptance of a monopoly of the airwaves. Reith was one of these men, and in large part the vision and reality of the early years of the BBC was a reflection of his substantial personality. He gave the BBC 'form and purpose', Asa Briggs has argued,[10] and others have gone further to say that the 'near absurdity of his vision enabled him to foresee the power of the new service'.[11] His megalomania and dominating style, steeped in Presbyterian morality, is well known, yet there was also thoughtfulness, personal torment and some sensitivity.[12] While serving with the 5th Scottish Rifles in 1915, he took care to introduce his men to God, distributing twenty-two 'little khaki-bound New Testaments'.[13] The sonic peculiarities of battle were remembered over forty years later in his autobiography: 'When one hears the vicious snap of a bullet the danger is past. The whine of a shell, pitch and volume according to nature and size, heralding its coming.' But Reith's excitement was evident too in his war-time diary: 'Most thrilling to hear the shells whistling through the air and to wonder how near they're going to land'.[14] The other men with whom Reith worked in the first BBC office on Kingsway in London lived with their own war experiences, which are relevant to the nightingale episode, as we shall see.

Reith was determined for broadcasting to be accepted as an essential part of national culture, if not of civilization.[15] Any suggestion that what rapidly became seen by many as the 'broadcasting craze' was yet another ubiquitous modern noise to accompany telephony, the gramophone, motor transport, train travel and uncontrolled outdoor advertising, was challenged by Reith with gusto.[16] At the same time, he wanted to see that 'any and all' Britons could have access to the world of politics and culture, which did indeed require ubiquitous reach across the nation.[17] In early 1924 one wireless magazine journalist delighted in the thought that 'the time is at hand when no place in forest, mountain or moor shall be too isolated to be linked with the life that is throbbing in the metropolis'.[18] Wilfred Whitten, editor of the popular literary magazine *John O' London's Weekly*, couldn't think of anything more intrusive. He described broadcasting as 'this immense new lure of life', but he was appalled by the promise of wireless everywhere because it would defile Romantic tradition and culture by invading the fresh air of the cherished English countryside. 'I wonder what Wordsworth would have said or done had he lived to know that the air of his Lakeland mountains and valleys was alive with our West-end tumult', he wrote.[19]

While Whitten was happy to see the loneliness relieved by wireless of far-flung lighthouse-men or Highland shepherds, he wondered why the sounds of nature could not be beamed into urban homes rather than voices from the city being relayed across Britain. In an article of March 1924 entitled 'The Lure and Fear of Broadcasting', Whitten lobbied the BBC:

I wish there could be an exchange of experiences between the silences of Nature and the hum of the city. I would set up my aerial to-morrow if, in the heart of London, I could hear the cattle lowing on the remote hills, or the barking of a fox in Essex, or the scream of an eagle over a Scottish glen. I would gladly summon the roar of Niagara to redress the roar of the Strand; but such things are not yet.[20]

Reith wrote promptly to tell *John O' London's* readers that the 'sounds of the country' had always been part of the broadcasting plan, and revealed that soon the 'liquid notes of the nightingales shall be borne in mystic ether waves to the home of the jaded town-dweller'.[21] Reith had indeed been planning with his engineering team since late 1923 the possibility of transmitting the live broadcast from a Surrey garden of a duet between a nightingale and the cellist Beatrice Harrison playing the Irish folk melody 'The Londonderry Air', the musical setting for the hugely popular wartime song 'Danny Boy'.[22] This would have been the first live broadcast from an outdoor location outside of London, and for Reith it needed to have more purpose and substance than a mere publicity stunt. Harrison, a family friend of Elgar and an expert with his cello repertoire, had telephoned Reith to persuade him, but she reports he was 'very dubious'.[23] Nevertheless, Reith was eventually convinced and sanctioned the experimental broadcast from Harrison's wooded garden for May, when the nightingale would, in theory, be in full song.

Eighteen months after BBC broadcasting had begun, at a quarter to eleven in the evening on Monday 19th of May, the Savoy Orpheans dance band was interrupted by the nightingale in full song. Studio engineers had been waiting for the bird to get going before switching over.[24] We should try to imagine the contrast between the foxtrot beats and jerky jazz syncopations coming from the Savoy Hotel ballroom in the Empire's capital, and the soft tones of the cello and the bird in the dark of the Surrey woods. This was a special moment on air, sufficiently so that live transmissions were repeated over several nights in May, the bird sometimes performing for fifteen minutes.

The broadcasts were a sensation, perhaps a million people listening in late at night.[25] Newspapers around Britain covered the event and the wireless press gave it front page attention (Figure 3.1). Harrison claimed to have received fifty thousand letters: an analysis of the few surviving ones written by men, women, the young and the elderly from the archives of the Royal College of Music makes it clear that for most the bird was the star of the performance.[26] Letters came from Huddersfield, Gosforth in Northumberland, Glasgow, and Belfast, regions well to the north of the nightingale's range. Through folklore, its singing abilities were legendary, and the nightingale had provoked more poetry in the English language than any other bird.[27] But few Britons had ever witnessed the song of this

Figure 3.1 Beatrice Harrison poses with her cello on the left, and the microphone set-up is shown on the right.
(From *Popular Wireless and Wireless Review*, 31 May and 7 June, 1924).

small shy brown bird for themselves, even though men in the trenches had heard it and written about the experience. R. M. Monk from Bramhall in Cheshire said this: 'I wonder do you know what it means to dwellers in the commercial north to enjoy for a few moments the pleasure of the nightingale's song—if you do, then all your efforts are rewarded.'[28] A man in Godalming wrote to tell Harrison that from the loudspeaker in his garden the broadcast nightingale provoked another of the species to sing along. W. J. Daully had also been enthralled:

> Will you please accept the very grateful thanks of a Liverpool postman and his mother for the great joy that you were instrumental in bringing to their ears last night. The reading [of] Keats' Ode before your work made one prepared for the atmosphere and mood and we heard the bird's notes very distinctly so too was your playing of the cello. By the same post I am sending you our leading morning paper and you will see by the enclosed poster that Liverpool well for one day forgot tragedy, politics, cricket and horse racing. We are anxiously waiting for further recitals. May we have them please.[29]

There are few other live broadcast events from outside the studio to compare this to; broadcasts from theatres and music halls were largely banned from 1923 to June 1925 to preserve their audience revenues.[30] The first major outdoor broadcast was the Duke of York's wedding in April 1923, where microphones were carefully placed outside to enable listeners to hear the sounds of bells, horses, carriages, and cheering crowds.[31] Microphones were not allowed inside Westminster Abbey that day, so what was heard by listeners at home were the incidental street sounds that conveyed the mood in London. The marriage ceremony itself was not heard at all. As a broadcast experience, this was a quite different to the singular nightingale performance lasting just a few minutes, where the microphone was close up and the mood was intimate and mysterious, emerging from darkness, without ceremony.

Just how much of the nightingale's song could really be heard appears to have been highly variable, perhaps because of different broadcasting conditions around the country and the many kinds of receiving equipment in use, some of which was home-made. During this early period of broadcasting listening in was a hit-and-miss affair.[32] Most listening happened through simple earphones or headphones, not a loudspeaker, so there was a certain kind of personal connection in play. From the letters sent to Harrison, some may well have heard a clear representation of the nightingale's song. For others, it may have been enough to hear the suggestion of the bird in song being broadcast live through the darkness. The *idea* of the nightingale's song, collected in Surrey, carried through the air, and reconstituted in one's home, was sufficiently potent for many. Perhaps its emergence through the pops and oscillations lent more magic. But not everyone was happy with what they heard. Some picked up only radio 'atmospherics', or was that twigs cracking under foot or the wind in the trees?[33] There was disappointment in some newspapers, the *Bristol Times and Mirror* arguing that many 'experts . . . might never have guessed what the sound really was' had they not known beforehand.[34] Some Birmingham listeners found that broadcasting was powerless to communicate the 'exquisite richness and wonderful variety' of the song, which could only really work its charms on the listener in a 'moonlit glade'.[35] That stories began to circulate about fakery and deception involving BBC bird imitators suggests that the broadcasts were in fact rather convincing.[36] Reith discounted critics who said that a mediated and commodified 'tinned nightingale' would take away the appetite for encounters with the real thing.[37] He refuted that the broadcasts were standardizing existence by distributing the world's experiences to the doorstep of anyone and everyone; rather they were a call to the outdoors and a chance for those who did not have the opportunity to hear such a delight to share in it.[38]

For all this, Reith was thrilled by the public response. He wrote in the *Radio Times* after the broadcasts that the nightingale 'has swept the country . . . with a wave of something akin to emotionalism and the glamour of romance has flashed across the prosaic round of many a life'.[39] Reith's use of the word 'emotionalism' is curious and powerful, usually meaning during this period a tendency towards a state of nervous agitation or hysteria. It was not a frame of mind Reith would have wanted to create with his broadcasting as a matter of course, nor was 'the glamour of romance' but this language indicates how special he thought the broadcast was. Reith received letters of appreciation himself. One of the first was from the 'head of one of the great industrial undertakings of the country' and this is what it said: 'I have heard and seen a good many remarkable things in my life, but the most remarkable thing that has come within my experience was the broadcasting of the song of the nightingale last night.'[40] This meant a great deal to Reith. The response to the broadcast was far more than a confirmation of the appeal of romantic myths—the song of the nightingale had made an impression on great minds and Reith sensed too it was full of symbolic meaning for the scheme of broadcasting.

Broadcasting silence

A crucial document with which to explore Reith's thinking at this time, and a landmark in broadcasting history, is his manifesto *Broadcast over Britain*, in which he reflects on the first eighteen months of his work as managing director of the BBC. He used the book to defend the monopoly position he had been given and to express his philosophy of a public service broadcasting that should 'bring the best of everything into the greatest number of homes'.[41] It is in the final chapter, which Reith named 'In Touch with the Infinite', that he explains how important the sounds of nature are to this broadcasting vision.

> Among the great paradoxes of life come the companionship of solitude and the voice of silence. To men and women confined in the narrow streets of the great cities shall be brought many of the voices of Nature, calling them to the enjoyment of her myriad delights. There is some peculiar quality about certain sounds, since they may not be incompatible with the conditions of silence. Already we have broadcast a voice which few have opportunity of hearing for themselves. The song of the nightingale has been heard all over the country, on highland moors and in the tenements of great towns. Milton has said that when the nightingale sang, silence was pleased. So in the song of the nightingale we have broadcast something of the silence which all of us in this busy world unconsciously crave and urgently need.[42]

Reith says a great deal in this striking passage. Nature's voices are a national delight and treasure, he argues, a sentiment that had been heard from Stanley Baldwin in the same year when he spoke of the sounds of the countryside being emblematic of Englishness. But more importantly for Reith, birdsong constituted a kind of silence—it was compatible with silence, could become part of silence, inhabit it.[43] This was something that broadcast human voices and music could not achieve, and Reith was acutely sensitive to suggestions that broadcasting created yet more modern noise and unsettled minds.[44] The editor of *The Times* was reminded by the nightingale broadcast that a good many 'are not over-keen on listening night after night to ephemeralities' of the human voice and human music.[45] Public opinion about BBC output was not in short supply and there was always discontent about the variety on offer: 'there was not enough jazz; there was too much jazz; the drama was too exciting; the talks were too dull; there was not enough light comedy; there was too much symphonic music'.[46] Standing apart from all this, the nightingale broadcast had been a chance to show that broadcasting could transmit sounds that were extraordinary, out-of-this-world, noiseless. Ernest Pulbrook in his romanticized chronicle of country life identified the paradox of how natural sounds could exist happily within silence. 'All seems very still in the beech wood,' he wrote, 'but if we stand awhile and listen the air is full of sound, almost *noiseless sounds* as it were—the cooing of the wood pigeon, the hum of insects, the call of the chaffinch.'[47]

The sounds and rhythms of nature had always been part of national culture. Literature, poetry, and music of the nineteenth and early twentieth centuries reflected the long-held and sensual love affair with the English pastoral ideal and the nature it was made from.[48] Broadcasting, Reith argued, was a servant of culture, 'and culture has been called the study of perfection'.[49] I think he offered the nightingale's song as an example of such perfection and included such sounds in his proposition of the 'best of everything'. Drawing inspiration from Matthew Arnold's *Culture and Anarchy* (1869), Reith wanted his broadcasting of culture to encourage Britons to resist the trivial and 'relish the sublime'.[50] Like Arnold, Reith wanted culture to be classless, and his thoughts on the sounds of nature specifically point to the enjoyment to be had by ordinary men and women of Britain's cities.[51] It is important to point out that much of Reith's best-of-everything would alienate many who were not familiar with or educated to be receptive to such treats. The sounds of nature on the other hand were open to all, requiring no particular education to appreciate. Even though the nightingale could appeal to the many, it was not 'mass culture' because it was not mass-produced. Rather it was rare, noble, pure, exquisite, enduring, moral, transcendent, yet all could appreciate it.[52]

Reith centred his last chapter, 'In Touch with the Infinite', on what he hoped would be the spiritual potential of radio broadcasting. The silence of nature was the silence of the universe watched over by his God. The song of the nightingale became in Reith's formulation a connection to the spiritual, if not a representation of the divine voice. These surprising thoughts were not disconnected from the social realities of post-war Britain. The traumas of the war were still fresh in the mind and the silence of remembrance that the Armistice ritual had introduced had brought new power to the place of silence in everyday life. With this rite 'this generation has received an imperishable lesson in the beauty of Silence' and had come to understand the virtue of stillness, Wilfred Whitten claimed.[53] Quiet, as we have seen in the previous chapter, was not only something that shell-shocked men needed. Loud noises still threw people into panic. There must have been many ex-servicemen who felt, as Robert Graves did, that it would be years before they could 'face anything but a quiet country life'.[54] Reith, and the men who established the BBC with him—Arthur Burrows, Peter Eckersley, and Cecil Lewis—had all had roles in WWI themselves. Burrows had run a news service during the war and had seen enough to become a life-long pacifist.[55] All were likely to be seeking their own kinds of social and personal stability after the war, and reflected this in their decision-making.[56] In wider intellectual and scientific circles, the war had left a feeling of fear that Western civilization was facing a terminal crisis.[57]

Reith was anxious that 'the broadcasting of silence' was not forgotten too quickly, because he saw the nightingale broadcast as an important break from the 'traditional stolidity of our race'. The feelings that had poured forth from those who heard the nightingale were actually a precious disruption to humdrum preoccupations with 'the review of sundry divorce and murder cases now proceeding; to the traffic problems of London, and to the threatened collapse of various bridges'.[58] Reith continued:

> There are times when the traditional stolidity of our race gives way. The barriers of reserve are broken. Latent and normally disciplined emotionalism is revealed. For a little while a measure of sentimentality is unashamed. Then, of course, 'better feelings' assert themselves. Cultured restraint, tempered with a measure of cynicism, holds sway again. The trivial weakness of the moment is forgotten; equilibrium is restored.[59]

The silence had punctuated busy lives with a chance to *feel* for a few moments. If this was acknowledged and accepted, then the resumption of modern life could be coped with a little better. To ensure that the effect would be recreated and become part of the permanent culture of broadcasting itself, Reith went on to authorize the nightingale broadcasts every May for the next twelve years, until Beatrice Harrison moved house.[60] Then the bird would sing alone annually until 1942, when the microphones picked up a British bombing raid in progress and the transmission had to be aborted.[61]

What Reith had done was to give the nightingale broadcasts a distinct part in the definition of public service broadcasting—a broadcasting that could refresh and enchant everyone, tugging human hearts towards the infinite, catering to people's unconscious needs, not simply their immediate compulsions and desires. In 1943, Anthony Asquith included a re-enactment of the duet between Harrison and a nightingale in his propaganda comedy *The Demi-Paradise*. The film explored English character and values as seen through the eyes of a Russian inventor, played by Lawrence Olivier, and spelt out that the strength of Britain's traditions, not least the eccentric ones and the belief in duty and service, would win the war. In one scene, Harrison playing herself performed with an unseen (and this time, recorded) nightingale in the garden of a country mansion, while two BBC engineers busied themselves to put it out live on the radio in spite of a German bombing raid in progress. The 'radio public must never be disappointed, Blitz or no Blitz', the bemused Russian inventor is told.[62] As Harrison wrote in her memoirs, recalling this film, 'all was to go on as usual', the bird singing along with the cello and the BBC broadcasting the event to the nation.[63]

The mutuality of nature and radio

The appearance of birdsong on the radio reflected wider societal interests in nature, evidenced by appetites for children's animal stories, W. H. Hudson's literary pastoralism, and Parry's musical setting of Blake's 'Jerusalem', introduced to the public in 1916. Nature study was now part of the school curriculum in England, and if *Punch* was to be believed, 'feeding the birds' had been a national pastime since before the war.[64] There was increasing scientific interest in bird life too. Harry Witherby had founded the influential journal *British Birds* in 1907 and the post-war period saw a definable subculture of 'birdwatching', encouraged by Julian Huxley and Max Nicholson, gradually replace the Victorian passion for collecting eggs, nests, and skins.[65]

Around the time of the nightingale broadcast and in its wake, enthusiasm for detecting and transmitting live the sounds of nature was all the more vivid. The 'Grand Howl' by George, the London station director's dog, was talked about as much by *Children's Hour* listeners as the nightingale was among adults.[66] The BBC's engineering chief, Peter Eckersley, conducted experiments at London Zoo with a cumbersome 'wireless perambulator' of gear to transmit the calls of a 'laughing jackass' bird and a hyena,[67] the first zoo noises being broadcast for children in November 1924.[68] In January 1925 there was an extensive preview in the *Radio Times* of an experimental night-time broadcast from an undisclosed location described as 'one of the finest wildfowl rivers in England'. While it was not possible to say when the transmission would take place, the organizers promised: 'We are fated to stand on deck and take the cries of the birds and the sounds of their wings as we sail up and down the river to transmit to you.' Along with the 'regular lap, lap, lap of the water as it strikes the boat', listeners could expect to hear widgeon, wild goose, mallard, teal, shelduck, swan, curlew, golden plover, lapwing, redshanks, stint, snipe, black-headed gull, herring gull, and heron.[69] On top of these examples, in 1924 and 1925 the *Radio Times* listed and printed features on talks and series about more zoo animals, butterflies in winter, tips on canary care, the nesting behaviour of the cuckoo, bees and gardening, the habits of the fox, identification of birdsong, the possibility of animal wireless communication (particularly insect communication), the habits of the dormouse, and the joys of listening to the radio outside in the summer in the garden, on motoring excursions, at the beach, and when picnicking.

What is clear is that some of these broadcasts and features were full of fun and novelty, but most were quite serious endeavours to bring new broadcast experiences into listeners' homes. The talks were aimed at adults as much as children, and while we know little about the reception of BBC nature programmes, Reith does single out and distinguish this genre from talks about physics, astronomy and chemistry when he says in *Broadcast over Britain*: 'Introductions to the study of Natural History, the habits and the ways of familiar animals and birds and fishes, have proved intensely human in their appeal.'[70] For Reith, nature programmes on the radio provided educational value, but they also provided a sense of the marvellous. The broadcast of live animal sounds was imbued with surprise and drama. When Reith engaged E. Kay Robinson as the announcer of the May 1924 nightingale broadcasts it was not only for his track record as a popular nature writer (he had a column in the *Daily Mail* from 1903 to 1906) but also for his beliefs in nature as a link to God.[71]

Nature mediated by wireless broadcasting took on its own kind of magic, and by association, the medium was touched by a similar aura. The radio set could tame the sounds of the natural world so they could be brought into the home, domesticated yet all the more strange and wonderful. Before the nightingale broadcast, the world of wireless radio was abuzz with the potential for microphones, transmitters, and electromagnetic waves to reveal mysteries. Peter Eckersley anticipated that his team's technical expertise would be able to portray nature unadorned in the 'howls of owls, the raving of ravens, the chaff of the chaffinch, the grousing of the grouse and the wheezing of weasels'.[72] Such excitement was mirrored by advances in 'ultra-microphone' developments which brought the possibility of '"conversations" with ants and bees'.[73] The BBC had its own purpose-built microphone commissioned and installed in its Savoy Hill studio in 1923, the highly sensitive Marconi-Sykes magnetophone. The first BBC director of programmes, Arthur Burrows, wrote that 'the story of this microphone . . . is itself one of the romances of wireless'.[74] This kind of romance harked back to the century before when the development of the microphone was accompanied by wonderment about what this new device might reveal about the natural world, hitherto beyond the reaches of the ear. In 1878, William Preece, the electrician of the British Post Office, had declared about the microphone: 'I have heard myself the tramp of a little fly across a box with a tread almost as loud as that of a horse across a wooden bridge.'[75] And a *Spectator* report on this new device looked forward to the chance 'to hear the sap rise in the tree; to hear it rushing against small obstacles to its rise, as a brook rushes against stones in its path; to hear the bee suck honey from the flower; to hear the rush of blood through the smallest of blood-vessels.'[76]

In this excitement is the idea that nature's teeming energy could be revealed by the microphone and broadcast for all to hear for the first time. This meant that broadcasting was more than a simple transmitter of man-made culture. The nightingale's song had been cast by Reith as a moment in broadcasting where minds could tune out of what was known and to rest, reflect, and allow bigger thoughts and sensations to appear. Next is a development of these ideas, taking the lead again from Reith, who wished for broadcasting to complement human senses in revealing not just the sounds of birds, sap, and insects but the resonances of the universe itself.

In touch with cosmic harmony

John Reith was interested in more than the comforts of Britain's natural history and wild places. The vision of nature he wrote about was given a capital 'N' and it encompassed all of physical creation, a realm in which he situated the physics of broadcasting as a medium made of loudspeaker, electronics, and electromagnetic radio waves. Reith developed the idea that wireless broadcasting connected listeners' minds to the harmonies of the universe, nature in its grandest and most spiritual state. Radio was a conduit through which divine stillness and silence could be channelled to humans. These ideas may have emerged as part of an urge to attach a secular theology to broadcasting at a time when religion in Britain was declining, or they may have been seen as a way to normalize radio by associating it with the natural, and thereby dissipating the fears inherent in all new mass communication technologies. Reith's idea of universal connection was certainly energized by the thrill of broadcasting's potential to do public good.[77]

As the 1920s progressed, Reith's programme schedule began to feature the topics of physics and cosmology, together with biology and natural history. The natural world, so often understood at this time through the evolutionary biology of Charles Darwin, was complemented by the wider historical dimension of cosmological evolution, which had become part of public knowledge and imagination by the time the BBC began broadcasting.[78] For instance, before WWI the discovery of radioactivity and the electron were widely reported in popular media, and *Punch* regularly carried cartoons about radium and the electron as well as radio.[79] Peter Bowler points out that after the war, Albert Einstein's name was well known, while his visit to Britain in 1921 revived again press and public interest in his complex theoretical physics.[80] The physics of the very small jostled with the cosmology of the very large in the public mind. What all this meant was that conceptions of the natural physical world could be comforting and homely, extraordinary and unfamiliar, or mind-bendingly abstract, but in this mix the public were exposed to all sorts of ideas about the microcosm and macrocosm of their surroundings. Such ideas were not necessarily unsettling. The reassurances given by Einstein to the Archbishop of Canterbury that his theory had no implications for religion was a signal that science would cooperate with Christian doctrine as it so often had, and it was also a reminder that faith could become 'ever more cosmopolitan', as Leigh Eric Schmidt has put it, in the presence of rational ideology.[81]

The BBC was part of this cultural conversation. In the summer of 1924, Reith gave away his front-page *Radio Times* feature to a biologist whose work sought to reconcile science and religion. Professor J. Arthur Thomson was a Scot for whom nature was the scene of divine activity. He was probably the best-known popular science writer and lecturer in the country.[82] Thomson's radio programmes used the vocabulary of fascination: *Animals That Work Miracles*; *Wonders of Underground Life*; *Marvels of Bird Migration*; *The Drama of Animal Life*.[83] These were popular science programmes that also acknowledged the extraordinary and hitherto unknown domains of nature's operation. Thomson and other leading scientists, including Julian Huxley, were interested in Henri Bergson's ideas of a non-material force of 'creative evolution', with a potentially moral purpose.[84] Reith's *Radio Times* front page was promulgating ideas that he would have taken great interest in. The science and technology of radio broadcasting seemed to rely on non-material forces too and Reith was keen to link these to the spiritual if possible. His own writing made efforts to do this, but there were others who took up these themes.

At the BBC, Arthur Burrows imagined the possibility that through wireless vibrations 'new points of contact with other realms of creation' might be possible. 'What surprises may be in store on the other side of silence . . . the love-songs of butterflies?', he wondered.[85] But the poet Alfred Noyes went further when he spoke of broadcasting as a miracle medium that reached deep into the universe. In 1925 he wrote in a *Radio Times* piece called 'Radio and the Master-Secret', 'The churches are beginning to preach from their pulpits that the age of miracles is over, and that all miracles are myths, at the very moment when science itself has revealed the whole universe to be an everlasting miracle.' Radio waves, Noyes said, were as Wordsworth described, 'a spirit that . . . rolls through all things'.

Broadcasting was proof that 'the Supreme Power was in communication with every part of the universe'.[86] There is quite a lot to take in here. In essence, Noyes has made the claim that radio waves were a miraculous work of God and they carried his messages across the universe. Reith's thinking is here, though Noyes has expressed it more vividly than Reith, who was not prepared to go quite as far.

Reith's *Broadcast over Britain* indeed airs in public some of his deepest beliefs and hopes for wireless broadcasting. Looking again at his chapter 'In Touch with the Infinite', we find Reith elaborating his thought, or wish, that somehow wireless radio offered a chance for humans to make contact with the unbounded cosmos and in so doing unite humans:

> Broadcasting may help to show that mankind is a unity and that the mighty heritage, material, moral and spiritual, if meant for the good of any, is meant for the good of all, and this is conveyed in its operations. So our desire is that we may send broadcast through the ether, which is universal, the universality of all that is good in whatsoever line we may; and so all may receive without let of hindrance, and without encumbrance or care.[87]

In his pulpit tone, Reith puts forward something of his vision of broadcasting for all humans—that it is a unifying force of electromagnetic waves travelling through space, without limits or barriers, distributing goodness. But for this arrangement to work Reith introduced into his argument the ether, a substance that filled the universe, connecting all matter and making wireless accessible to all.

Connecting to harmony

The ether was a long-standing concept in scientific thought, but physicist Sir Oliver Lodge had popularized the idea of a substance that would 'weld' atoms together in space, transmitting vibrations from one piece of matter to another.[88] Lodge contended that wireless could only function in the presence of ether because this was the medium through which radio waves travelled. More ornately, the ether was central to the 'romance of wireless' because it is a 'vehicle for messages', as Lodge put it.[89] Lodge was a major public figure, a pioneer of wireless communication, the scientific advisor for a leading magazine for amateur wireless enthusiasts, *Popular Wireless*, and a charismatic writer and BBC broadcaster. In 1925 he hosted a seven-part radio series that started with 'The Mystery of the Ether', which featured on Reith's front page of the *Radio Times*.[90] Yet Lodge's ether theory had been largely undermined by Einsteinian physics, which had no need for an etheric substance for electromagnetic waves to act in the universe.

Reith would have been well aware of this; nevertheless, he and others found the metaphors wrapped up in the ether appealing and useful. 'Wireless is in particular league with the ether', Reith wrote, and he revelled in its mystery.[91] Ether was employed to explain telepathy, experiments about which were hosted by Reith and Lodge on the BBC in 1927.[92] Apart from being a medium that might allow minds to communicate silently, the invisible ether carried connotations of the clear blue sky, the heavens beyond, the pure essence breathed by the ancient Gods, and none of this was lost on Reith.

In *Talks about Wireless* Lodge wrote: 'the ether welds the worlds together into a cosmic system of law and order'. The ether made coherent what would otherwise be chaos. It resisted disorder to ensure the universe was held in a state of perfection. It provided poise and equilibrium. Lodge saw a social as well as a cosmic function: 'Let it weld all humanity together, so they can face their common difficulties in a spirit of cooperation and mutual trust!'[93] There is correspondence here between Reith and Lodge, in what they see as a moral purpose for radio in bringing the world closer together in peace.[94] Within these thoughts of global harmony can be found the ancient notion of the 'music of the spheres', where each celestial body contributed a unique tone to a great melodic assembly.[95] As Reith considered music to be an international language it is quite possible that he imagined broadcasting to be a method of relaying a kind of cosmic harmonic message.

The mystery of the ether and the mystery of broadcasting would not be deciphered by humans, Reith was certain. These things were quite beyond human senses and knowledge.[96] Following Lodge, Reith asserted that human senses were 'painfully inefficient' in detecting the 'vast ranges of vibrations with things happening that we cannot get in touch with'.[97] Indeed, the last page of *Broadcast over Britain* makes clear that these human shortcomings should be accepted, and that spiritual contemplation was the route to enlightenment:

> We should also be aware of the feebleness and errors of our own perceptions and intelligence, and from this awareness, turn to the contemplation of the Omnipotence holding all things together by word of power, in Whom, as in the ether, we live and move and have our being.[98]

This is as close as Reith comes to explicitly placing his God into the picture of broadcasting. Nevertheless, it is clear he believed that the BBC could be a public service that included a mystical dimension. The best of everything would include access to if not communication with the heavenly presence. The formulation that comes to light in this analysis of Reith's thinking can be summarized in a linear flow: God > Cosmos > Ether > Waves > Radio sets > Ears > Minds. This scheme of what I think many who were close to broadcasting had in mind shows that what was on offer was contact with the divine. This was not a two-way conversation. Listeners could receive but not send. The onus in this model was on the listener being open and receptive to the connection.

Following these ideas through, it would have been possible to believe that at the distant end of that connection was the bounty of heaven, part of which was the sound of silence, or perhaps the singing of angels. When Reith wrote about the significance of broadcasting the nightingale's song he had invoked the 'many voices of Nature' as bringing a special kind of 'silence' to the lives of his city-dwelling listeners. However, it seemed that the most pure and perfect silence was not the one that birdsong could inspire, but that associated with the infinite. There was nothing to fear in this immense silence; instead it might offer Reith's 'companionship of solitude'.[99] Though Lodge and many others sought an exchange through the ether with the dead after the war, Reith's infinite was not a place of hauntings.[100] The silence would have been associated with contemplation, transformation of the self, prayer, and connection with God. A silent creator overseeing a silent universe was a story of scripture and of Western culture.[101] Reith wanted to somehow draw that silence into his broadcasting system, an electronic medium of pops, whooshes, and puzzling groans though it was. Wilfred Whitten commended the rare commodity of silence in modern life to the readers of John O'London's Weekly. He said that people longed for 'the silence in which the greatest sounds can be heard more abundantly', especially 'the whispers both of earth and heaven'.[102] Stillness for him was a virtue, a magnificent hush that could allow better thoughts and feelings to appear.

The more prosaic Sunday schedule of religious observance was a terrestrial source of silence accessible more easily to everyday listeners. Reith located the most important part of Sunday broadcasting in a change of pace, as much as a change in content. The rhythm needed to be stepped down in tempo and there needed to be less on air to allow citizens to attend church services. There were no transmissions at all during 'Church Hours' on Sunday morning, just two hours of music in the afternoon, and 'then nothing till eight or half-past' in the evening when there was a 'short service sent out from all the studios', or every month or so a complete church service that carried the atmosphere of the church, including its bells.[103] Broadcasting was in large part silenced. Wireless sets could be switched off.[104] Reith summarized the effect of this carefully limited day on the air in this way:

> Apart from any puritanical nonsense, I believe that Sabbaths should be one of the invaluable assets of our existence—'quiet islands on the tossing sea of life'. It only requires a little thought to determine how best they may be employed, and how turned to greatest advantage. This is not to be achieved by sport or motoring or parading about the streets. It is a sad reflection on human intelligence if recreation is only to be found in the distractions of

excitement—if no provisions are to be made for re-creation of the mind and refreshment of the spirit; the spirit is surely of at least as much moment as the body, and many of the ills of the latter are attributable to the neglect of the former.[105]

It seems that Reith acknowledged that some may not benefit from strict religious protocols and messages, but all needed and enjoyed a break from work-a-day routines, including listening to the wireless. Peace and contentment could be found in the communion with the right quiet pursuits, although we are left to wonder what exactly Reith would prescribe if he could.[106]

Normalizing radio with nature

While Reith made efforts to spiritualize the medium, there were at the same time wider efforts by the BBC to normalize radio by associating it with nature and thereby anchoring broadcasting in the known. There was a balancing act going on here, though how consciously is unclear, in that broadcasting was cast as simultaneously transcendent and a part of the everyday. Broadcasting could be both. However, the normalization of radio by associating it with the natural world was not simply a matter of relating the medium to everyday terrestrial nature, though this happened; radio was also linked to the powerful atmospheric forces surrounding the earth.

Director of programmes, Arthur Burrows, like Reith, published an account of the first eighteen months of broadcasting. In his preface to *The Story of Broadcasting* Burrows was keen straightaway to associate broadcasting directly with the planet's energy systems:

Nature has been 'broadcasting' since the earliest thunderstorm. With the first lightning flash, wireless waves were sent rippling across space, penetrating primaeval forests, rocky caverns and the haunts of such animal life as then existed.

Man himself has been broadcasting for over a quarter of a century, using harnessed forces of nature and the wonderful discoveries of Senatore Marconi and other workers in the same field. These wireless waves, like those born of the earliest thunderstorm, have been passing quietly though our homes and our bodies. To us they have meant nothing.

The wireless station from which these waves originated has been viewed as a mystic place, having less bearing on the affairs of the man in the street than an astronomical observatory.

To-day the position is entirely changed [. . .] every other person is now discussing broadcasting or some item in the broadcast programmes.[107]

Here Burrows suggests that broadcasting through the ether is an entirely benign process that nature itself has always been doing. There was certainly nothing to fear from wireless radio waves 'passing quietly through' homes and bodies, now an everyday event. Burrows deferred to nature's comforting age-old processes to defuse any anxieties that wireless broadcasting might be a threat to human bodies and minds. This kind of narrative was a response to, Burrows freely admitted, 'the quite common case of persons complaining that broadcasting was injuring their health'. Even if electromagnetic waves were part of the way the earth's systems worked, man-made radio waves were cause for concern. Some had reported that birds were seen to drop dead in their hundreds when flying in line with wireless waves.[108] *Popular Wireless* too had reported, light-heartedly, that gulls and doves could lose their direction-finding senses because of 'some effect of the ether waves not yet understood'.[109] There was speculation that the ever-increasing use of wireless broadcasting might affect weather conditions and climate more generally, through its disturbance of the ether.[110] It was small wonder that these concerns emerged, because the knowledge that wireless waves did indeed pulse across the planet and through the universe at the unimaginable speed of light gave pause for thought to anyone who pondered such things.

These anxieties were countered enthusiastically by the BBC and others, not by arguing that wireless was safe, but by insisting that wireless was good for you.[111] *Radio Times* and *Popular Wireless* bustled with articles and letters about 'wireless healing'. Children listening to wireless were 'more contented and robust', perhaps through 'atomic electricity . . . picking up some unknown force in its transmission on the ether waves and delivering them to the recipient', one writer speculated.[112] A medical correspondent told of the 'life-giving wave-lengths' that improved colour and boosted the spirits by direct action on the central nervous system reached through listener's headphones.[113] Wireless listening offered a 'rest cure' for the run-down others suggested.[114] In the summer of 1925, the BBC engaged several authoritative voices to get the message across in the *Radio Times*. Sir Bruce Bruce-Porter, KBE, CMG, MD was given the front page for his reassuring article 'Health and Headphones',[115] and Lord Knutsford in 'Wireless for the Wards' announced a new initiative to get wireless installed at the bedside in all London hospitals.[116] If radio waves were part of the normal workings of nature, then it could be argued that wireless energy was not a threat but health-giving. Reith and Burrows had in their own ways facilitated the construction of stories about the benevolent psychic architecture of broadcasting, led by their enthusiasm for the new venture as well as the need to legitimate the service with the public.

Many others took part in the promotion of the notion of wireless waves as part of nature. 'What is more *natural* than wireless?' asked a writer in the up-market magazine *The Broadcaster*. He continued by adding that 'sight, sound, heat and light are all wireless'.[117] Staff writer J. F. Corrigan, in his May 1926 feature 'Aerials in Miniature', drew direct parallels between 'ether vibrations' set up by wireless broadcasts and those by 'insect wireless'. Radio science had potentially answered how insects communicated with each other using ether waves transmitted and received though their antennae or 'feelers'.[118] An intriguing feature article appeared in *Popular Wireless* in 1923 that showed the enthusiast how to make a loudspeaker from a seashell (Figure 3.2).[119] Radio enthusiasts may be 'essentially modernists', the piece said, but they would do well to experiment with one of nature's loudspeakers—'the despised ornament of the Victorian home is really an ideal component for the most scientific instrument'. The guide showed how to drill out and fit a headphone earpiece and as a result the seashell as a loudspeaker was found to be 'remarkable for its mellow tone and clear reproduction of both music and speech'. That a pleasant sound should emerge from a shell that had previously had the eternal murmur of the ocean playing inside it seemed logical, in a strange though not-very-scientific way. There was also a pleasing symmetry in wireless's natural waves being shaped and relayed to listeners through the curvaceous aperture of a seashell, whose aesthetics, in turn, off-set the harsh science lab appearance of the receiving apparatus. Perhaps the message was that radio people appreciated that nature could be complementary to progressive modern ideas.

Thinking of wireless broadcasting as part of nature lent the medium the enchantments that were found in the natural world, great and small. Burrows found himself ruminating on the molecules, atoms and electrons that made up a single morning dewdrop. He looked to the heavens on a cloudless night, 'across a boundless space, to other worlds, each one of which, suspended by an invisible agency and spinning like a golden top, pursues in silent motion an endless well-defined course'.[120] Here is the peace and stillness that Reith perceived, Burrows noting its perfection undisturbed by wireless waves and vibrations. Here is the orderly pattern of nature, bounded by its own laws that broadcasting was itself built upon. When Burrows talks of the 'mysterious' ether, he is talking of a beauty too great for humans to comprehend. The ether is part of a natural order of planetary movements, earthly season cycles, fluxes of light and darkness, tidal forces, and sap rising and falling.[121] In Burrows's account of the first eighteen months of BBC broadcasting, wireless is depicted as something natural but at the same time part of the greater mystery of the cosmic infinite and all the more powerful and exciting for it. Broadcasting was extraordinary, like so many other natural forces of creation.

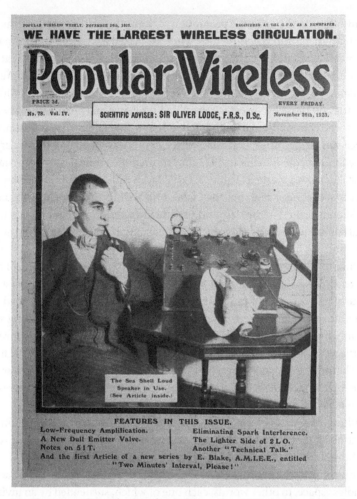

Figure 3.2 The cover of *Popular Wireless* in November 1923 showing how a large seashell could be rigged-up to act as a loudspeaker.

Listening outdoors, in nature

In the early years of the BBC's work, broadcasting was also normalized by promoting listening as a complement to popular outdoor pursuits for women and men. During the first summers of BBC broadcasting, the *Radio Times* and wireless magazines were taken with the idea that wireless listening could happen anywhere (Figure 3.3). It was quite possible, readers were informed, to take the wireless with you—out into the garden, on picnics, or to the seaside—to enjoy the glories of the summer while keeping in touch with the

Radio in the Summertime.

SOME OF THE JOYS OF OPEN-AIR LISTENING.

Figure 3.3 Some of the possibilities for listening outdoors suggested by the *Radio Times* in May 1925.

news or a concert.[122] Wireless set manufacturers conjured the appeal of listening to the wireless while driving or boating. While some of this was certainly a marketing effort (for the BBC to keep listeners listening and for set manufacturers to keep receivers moving out of the shops),[123] there was more at stake.

The Broadcaster columnist Russell Mallinson wrote about all kinds of people listening outdoors—the many who might have a cheap and compact crystal set, or the more well-off who might own a bulky valve machine that needed heavy batteries.[124] Valve sets were hardly portable and would need a motor car to be able to get them out into the countryside.[125] Modern suburban families had to be catered for by broadcasting and they were part of the imagery of radio listening within summertime nature.[126] Radio could provide a social glue for the family: 'The programme has to get into the home atmosphere', wrote Cecil Lewis, who worked alongside Arthur Burrows.[127] It was imagined that homely family normality helped along by the radio could be reconstituted by the seaside or under the shade of a tree with a picnic. Western Electric ran a series of glamorous full-page advertisements showing their 'wireless receiving apparatus' aboard a yacht under sail, as part of a lazy river punt, and in the garden beside a game of tennis. These were fantasies of a mobile future where wireless was always to hand, projections of the happy habit of wireless listening. They were also a reflection of the fashionable status of listening and being *seen* to be listening in, a new public performance.[128]

Listening to the radio in the countryside was part of a broad social communion with the landscape that was gaining pace at this time. Camping, youth hostelling, and rambling were notable activities that made use of the countryside for leisure, pleasure, and an assertion of English identity.[129] Camping after the war was seen as a healthy outdoor pursuit of the type needed for 'invigorating and healing jaded minds and bodies'.[130] Radios were not central to camping culture by any means; in fact many people were seeking to escape the mechanized habits of urban life. Yet there were reports of charabanc trips into the countryside where travellers would be entertained with music from the on-board wireless set: 'In Lancashire these juggernauts of the highway are in many cases fitted with radio receiving sets, which through the medium of a loud speaker, broadcast concerts to the passengers during country runs'.[131] In such collisions of nature excursion and wireless listening can be seen ideas about the romance of the English rural scene matched with the romance of radio, the latter so often invoked in the first years of broadcasting. Reith had commented on the romance of radio when explaining the effect of the nightingale broadcast on the nation, as so many journalists did;[132] Arthur Burrows talked about the microphone as a romance of wireless; Oliver Lodge had used the term when writing about the ether; and department store magnate Harry Gordon Selfridge announced in *Popular Wireless* that 'romance, after all, is the spice of life, and in the radiophone receiver we have romance personified'.[133] The wonderment of British nature and the glamour of the new wireless world seemed to go well together.

The paradox of listening that this conjunction created could not be avoided, however. This was because whether listening to the wireless outdoors with headphones or with a loudspeaker, both modes obliterated the possibility of engaging with nature's own soundtrack. Although it may have been the case that having nature at one's disposal while being immersed in the modern mode of wireless participation, in public, was the best of both worlds. The journalist Sinclair Russell saw wireless music as entirely complementary to the mood of the rural scene. There could be dancing by day and by moonlight, and he predicted that 'the breezy expanses of the Norfolk Broads this summer will re-sound to the echoes of loud speakers on yachts and wherries'.[134] Another writer was charmed by the prospect of wireless music for 'healthy recreation' from the 'turmoil of the cities', 'challenging the night owls amidst the verdant beauties of some forest clearing'.[135] This unlikely drama was explained by a belief that nature's own green acoustics were particularly flattering to many sounds, including wireless music:

> For the silence of a garden brings a new charm to music and voices which makes one forget the intricacies of the science of acoustics. One might as well adapt such theories to the song of a lark or the warbling notes of a nightingale. And here perhaps nature has created a scheme of her own. For it may be that the grass, the foliage and the trees have a subtle power to enhance the beauty of music with acoustics that nature has provided.[136]

In this imagining, wireless had been completely accommodated by the natural world. Broadcasting belonged in nature because it sounded better there. There was no reason why the song of the lark could not be heard in harmony with wireless sounds. After all, the nightingale's song had been brought by wireless into homes around Britain; the sound of the country brought into the city. Why then should the sounds of man and his music not be sent into the countryside to mingle with the sounds of nature? This mutual exchange apparently dissolved the tensions associated with ubiquitous wireless listening and placed the new medium comfortably within home and outdoor listening arenas as a national instrument of public service.

Conclusion

The very first broadcast sounds that emerged from the wireless in 1922 were heard to be nature calling out from the ether. 'Many people first heard the wild waves calling through the earphones of a crystal set', Paddy Scannell and David Cardiff have said in their social history of broadcasting.[137] Before the live broadcast of a nightingale singing with Beatrice Harrison's cello from the Surrey woods, the airwaves were already alive with aerial mysteries. The sounds of the air, of wind, of water rushing as well as the uncanny murmurs of spirits were the first to be recognized by expectant listeners. The sounds of the natural and the supernatural mingled. The nightingale broadcast is well known still today, though little explored in scholarly work, and it has resurfaced in recent years to be portrayed by the BBC as a charming and idiosyncratic moment in its history. But it is much more than this. In the establishment of the BBC's public service, the sound of the nightingale on the wireless was at once symbolic of the enchantment that modern life was missing, but that the technology of broadcasting could supply, as well as part of the scheme John Reith gradually evolved to bring a spiritual dimension to both the content and medium of broadcasting. If few could agree what should be on the wireless, most seemed to feel that a nightingale would be a good idea from time to time.

Before the BBC's activities brought the term into wide circulation, 'broadcasting' connoted the cultivation of nature by the scattering of seed, a metaphor that Reith brought into social practice as the free distribution of culture worth nurturing and propagating further.[138] One commentator wrote that Reith could be thought of as a 'gentleman farmer' in his work as a public service broadcaster. The relationship between the new cultures of radio and the ancient ones of cultivating the natural world were remarkably close in the early years of the BBC. Reith wanted his broadcasting of culture to go beyond human talking and musicking and his religious commitments demanded more of him. He found in the transmission of the iconic sound of the nightingale in song a sublime signal from deep within Britain's natural heritage. It was part of the promise of the best of everything, which needed surprising revelations from time to time that appealed to all classes and tastes. Paradoxically, Reith could claim that his nightingale communicated a silence craved by urban dwellers all over Britain, even though it was barely a nightingale by the time it crackled into homes. At least the bird's song could represent a silence or a comforting stillness or a pause that could counter the rush of modern lives, brittle in the wake of war. If Reith's broadcasting was a tool for improving modern life, then engagement with British nature provided the means to cope with it. The combination was a potent one.

Along with scientists and cultural critics, Reith believed that the medium of radio itself was shot through with mystical potentialities. Radio waves and the mysterious ether offered connection with cosmic nature and even the heavens, domains of stillness or the music of angels. The possibility that the new broadcasting medium had an intrinsic celestial resonance within its network was deeply appealing to the men at the BBC. It meant that, like Reith's Sabbath-day programmes—'quiet islands on the tossing sea'—broadcasting could offer transportation of the soul.

4

The rambler's search for the sensuous

Not to find one's way in a city may well be uninteresting and banal.
It requires ignorance—nothing more. But to lose oneself in a
city—as one loses oneself in a forest—this calls for quite a different
schooling. Then, signboards and street names, passers-by, roofs,
kiosks, and bars must speak to the wanderer like a twig snapping
under his feet in the forest, like the startling call of a bittern in the
distance, like the sudden stillness of a clearing with a lily standing
erect at its centre.

Walter Benjamin, 'A Berlin Chronicle' (1932)[1]

This chapter explores the ways in which walkers in the 1920s and 30s chose to
abandon themselves in the natural world at the weekends (and sometimes over
longer periods) in order to escape the standard sensory repertoire of urban and
suburban life. The character of sonic experience in Britain's rural areas will be
examined within a conception of all senses working together. Scholars generally
agree that there is no such thing as singular raw sensation, rather that the senses
operate with the body and mind working together—nerve cells, cognitive, affec-
tive, motivational, and societal processes in complex interaction—yet many opt
to study the senses individually.[2] Here I will account for several senses at once,
adding to hearing and seeing that have been key so far in this book the sense of
touch in particular, for many accounts of walking and being in the rural elevate
this sense. What emerges with this approach of taking account of other senses
is the appeal of the sensuousness of nature. Listening and touch work especially
well in close proximity to nature, allowing intimate connection.[3] Ramblers
sought out sensual experience that came to them through the combined effect of
an ensemble of multisensory encounters.[4]

Rambling was a primary practice of engagement with nature in the interwar
period. It was often spoken about as an escape from the trials of the urban and
industrial worlds where more and more people lived out their lives. In that respect the escape was a removal of the modern persona from the new sensory encounters with machinery and sonic media and from the systems that

Listening to British Nature. Michael Guida, Oxford University Press. © Oxford University Press 2022.
DOI: 10.1093/oso/9780190085537.003.0005

regulated and constrained movement and sensation. City sensation might be overwhelming but it might also be impoverished though the confines and monotony of domestic work. A rhythmic weekend exchange from town to country and back again was driven by a search for tranquillity, but equally a search for stimulation and alternative sensations. Any notion of escape could only be temporary and can be seen as a way to *balance out* a range of sensory stimuli that modern living could offer. In this respect, the relationship was complementary more than antagonistic.[5] Rambling, then, enabled an exchange that millions enjoyed either as informal groups of friends, with organized rambling clubs, or as part of youth hostelling and camping trends that boomed in popularity in the 1930s. Cyclists were part of the weekend exchange but a feeling of being at one with the land was 'denied to the cyclist insulated by his tyres'.[6] Rambling was a way to connect to the earth.

Two debates have dominated scholarly work about rambling as part of the wave of interest in interwar outdoor leisure. The first concentrates on the political struggle to gain access to the hills, keep footpaths open, and win the freedom to roam. This is a key story of the changing British relationship with the natural world and one led by working class men seeking self-improvement and independence.[7] The second examines the more insular debates about the preservation and correct use of rural heritage seemingly under attack by the masses propelled by cheap trains and buses.[8] However, both of these struggles were predicated on the common urge to experience the sensory pleasures of moving about on foot through moors and mountains and along riverbanks. Without this desire for the sensory rewards that nature could provide there would have been no rambling clubs or boisterous charabanc trips, and no fights for access to the peaks of Derbyshire or protective rural codes of practice. This chapter exposes the underlying everyday passions that motivated ramblers of all persuasions, including those merrymakers who seemed to prefer to listen to music than their natural surroundings.

Diaries, journals, guide books, and newspaper columns bring to prominence the fondness in the 'townsman' for 'scenery' and 'beauty spots', very much terms of the time, but these words reflected the enduring trope of the Romantic gaze in British culture and can be interpreted, as well, as shorthand for a deeper-than-visual longing for close connection to natural environments that stimulated the whole body.[9] The ramblers' search for the sensuous included experiencing the sensations of bad weather and the thrill of dipping the body into a chilly upland lake. We will hear of midnight rambles on the Sussex Downs and winter walks high on the Derbyshire fells. The happy rhythm of rambling songs will contrast with contemplation of the inner world of the Scottish Cairngorm mountains.

Re-Balancing the senses

Rambling was and still is the practice of rural walking for pleasure. Though it is associated with the formation of the Ramblers' Association in 1935, it is most firmly rooted in a turn-of-the-century outdoor activism led by characters like G. H. B. Ward, who formed the influential Sheffield Clarion Ramblers in 1900, naming them after Robert Blatchford's socialist *Clarion* newspaper. Ward's working-class ramblers tackled long distances on the moors in all seasons. Before this, walking for pleasure became a noted feature of British culture with William Wordsworth's explorations of the Lake District, emerging more widely as a mid-Victorian gentleman's game. Leslie Stephen's walking group called the Sunday Tramps began in the 1880s and comprised a group of friends including the novelist and poet George Meredith.[10] Taken together, tramping and wayfaring, as well as rambling and hiking, were all ways of walking for a wide social spectrum between the wars. Manual workers, shop keepers, clerks and civil servants, small businessmen, and the many unemployed were drawn out of the towns and cities. Couples walked, and young women walked together and alone.[11] Many wrote of a rambling and hiking 'craze', Robert Graves and Alan Hodge recalling that 'hiking began to enjoy a boom in 1931'.[12]

Social commentators and ramblers themselves made it very clear that rambling was a way to escape urban life. The words of John Keats' poem 'To One Who Has Been Long in City Pent' ring out in much writing about rambling in the North of England. The cinema and the pub offered regular escape for the mind, but these were still urban, as well as indoor and sedentary, and they could only relax the body not enliven it. The cultural critic, rural preservationist, and moral arbiter C. E. M. Joad told a rambler's demonstration in Manchester in June 1933 how people used to 'get drunk to escape the squalor and misery of the towns'. Today, rambling in the most beautiful countryside in the world 'was the best way' he told his audience.[13] When J. B. Priestley's *English Journey* took the national temperature in the depths of the Slump, it set industrial and urban life in opposition to the rural. Rambling was a means to 'escape the conditions [. . .] of apathy and despair' induced by life in industrial cities like Glasgow.[14] Rambling clubs offered an alternative based on simple living and a reverence of nature, in the tradition of John Ruskin and William Morris.[15] Arthur Leonard, chairman of the Co-Operative Holiday Association that provided out-door fellowship for young people including walking holidays, felt that his was one group of many 'all over the country who were trying to stem the tide of mechanicalism'.[16]

The apparent sensory overload of city life that occupies much of today's cultural analysis of the early century often settles on Georg Simmel's writing about new and unexpected sensory experiences of the European metropolis. His compelling idea that urbanites developed a blasé attitude to their surroundings in the face of overwhelming and rapid sense impressions suggested an incapacity to respond to these stimuli. If urbanized senses no longer functioned in full, they were also shielded by the blasé attitude and techniques of avoidance. Rambling in interwar Britain was one way to distance oneself from the noise, smells, and speed that were said to disturb everyday life.[17] If the senses in the city were bombarded, they were also *regulated* by rigid work and domestic routines, new traffic and pedestrian controls, and the standardization of the emerging suburban grid. Ramblers were galvanized into action by the bland monotony of the towns and cities as well as their overbearing demands. Could rambling stimulate into excitation the dulled and bored urban body?[18]

The story of escape, then, is better thought of as a search for alternative sensory experience. As most rambling excursions happened at the weekend, we can picture a pattern of rhythmic exchange between home or work and the natural world outside of these confines. The escape was a balancing out of urban responsibility and rural refreshment. Mark Jackson's work on scientific and social concepts of stress highlights the circulation of ideas in the twentieth century to do with a search for stability and equilibrium in modern life. Anxious individuals, he argues, were the product of societies characterized by industrial unrest and social disequilibrium. Jackson points out that the concept of homeostasis, as well as a human tendency to obey physiological and psychological laws, was a political idea.[19] In rambling, people may have found a method of managing their senses, their physiology, their psyche as part of a wider effort to regulate the stresses of modern life for themselves.

An instinct to keep the senses fed and in balance emerged early in life. In her tiny Charles Letts diary, Catherine Gayler kept a neat daily record of the year 1934.[20] She was a diligent schoolgirl of perhaps fifteen years of age from the village of Carlton Scroop, near Grantham in Lincolnshire. The life of a child at school, grappling with homework and *Jane Eyre* and *Tarka the Otter*, is punctuated by her frequent outings, often up to the heath. She finds a hedgehog and takes it to school, gathers pussy willow, gets 'very close to a hare nibbling corn in a field', does a night walk along the beck, sees a nest of chaffinches by the bridge, and in her holidays hikes or cycles long distances every day. All of these small but notable sensual pleasures are found through the habit of 'hiking', as she calls it, away from home, into the natural world, and back again. Observations like these, alongside 'learning about the moon in geog' at school and exam 'swotting' make up the pattern of Catherine's early life.

Keeping the sensing body fed and in balance was formalized in the establishment of rambling clubs. G. H. B. Ward was a power station fitter who saw his work in forming the Sheffield Clarion Ramblers as a way to help men develop 'character' beyond the confines of work life. His idea was to encourage men like him 'to go during the weekends and enjoy the beautiful scenery around Sheffield, and, in this way, get rid of many of the ills and rheums which accumulate largely because men do not know how to throw them off in the abandon with nature.'[21] Ward was not particularly interested in standing around looking at the view—his ramblers got stuck into and on top of the exposed heather moorland that could be glimpsed from the streets of Sheffield. There lay freedom from 'the prisons of economic necessity', which cramped the spirit as well as the limbs.[22] The chorus of Ewan MacColl's song 'The Manchester Rambler' spelt out this sentiment and enlivened it with music: 'I may be a wage slave on Monday / But I am a free man on Sunday'.[23] Indeed, the Kinder Scout trespass in Derbyshire in 1932 and the many unreported incidents before it were driven by a fundamental belief in equal access to free immersion in all corners of the natural world. The ensuing debates about the right to roam can be seen as a bid for the right to rural sensory connection for ordinary workers. Ramblers were not looking for a fight with game-keepers—'what they wanted was respite from the "push and pull" of factory and city life'.[24]

In the bigger cities of Glasgow and London, for instance, the incessant jangle of machine noise was said to have propelled office and shop workers out into the countryside. The *Ramblers' Handbook* of the London Federation suggested that the newly formed Youth Hotels Association, along with rambling clubs, 'opens the way to the sea and the hills for those who know only the smell of the streets and the noise of the traffic'.[25] Rambling was a way that 'the pedestrian may still escape from the horns of motor cars and the syrens of charabancs' argued the Glaswegian George Eyre-Todd in his foreword to *Citizen Rambles*.[26] Traffic was not the only problem. Early twentieth-century sound media, especially the gramophone and radio, marked 'perhaps the most radical of all sensory reorganisations in modernity' the media historian John Durham Peters has argued.[27] Sonic media made new demands on perception and placed a burden of training on the senses. The ubiquitous sound of radio was too much for George Orwell: 'There are now millions of people, and they are increasing every year, to whom the blaring of a radio is not only a more acceptable but a more normal background to their thoughts than the lowing of cattle or the song of birds.'[28] Direct experience of life was being lost.

Rambling, however, would put you 'beyond the reach of letters and telegrams and telephones', Arnold Haultain wrote in his philosophy of walking.[29] After an autumn walk in Buckinghamshire he sketched this scene: 'A great mist lay over the land; a gentle noiseless mist that hid you from the horizon and the outer world; that shut you in from the outer world; that lured you into that mood of quiet reverence in the presence of quiet wonder-working Nature; and revealed to you . . . I cannot tell all that was revealed.'[30] This kind of stillness was cherished by the rambler, though few appreciated that sometimes it was a result of the depression in the rural economy. Many farmers faced bankruptcy in the 1930s and left fields uncultivated and hedgerows unkempt.[31] Nevertheless, the quiet matched with expectations and it reflected an idea that the British, or at least the English, were themselves a quiet people. In his 1928 fundraising appeal for the Council for the Preservation of Rural England, Lord Crawford argued that the natural heritage of the land 'reflects our national character—it has qualities of calm and quiet and intimacy'.[32] Such rhetoric played off anxieties that the peace of the rural was being gradually eaten away by the 'sprawl' of new suburbs with their ugly uniform geometries, overhead power cables, and a little car waiting outside.[33] The clutter and clamour of modernizing Britain was often associated with the careless driver and the brightly painted signs that were put up to attract his attention to chocolate advertisements, the services of the A. A., or a new tearoom.[34]

Motor traffic in British cities increasingly determined pedestrian movements and rhythms. Pedestrian continuities of flow were broken up by the challenge of crossing the road, which was made up of fast- and slow-moving private cars, taxis, motorcycles, double-decker buses, and heavy horse-powered wagons. The street was less and less the heterogenous social space it was after the war as its purpose became redefined as a functional conduit to speed the flow of traffic.[35] As crossing the street became dangerous, the public were asked to sharpen up their perceptions in order to cultivate 'road sense', and the Transport Minister Leslie Hore-Belisha by the mid-1930s attempted to orchestrate the behaviour of drivers and pedestrians with speed limits, horn-sounding restrictions, and controlled crossing points for walkers.[36] Attempts to smooth the asynchronies of stop-and-start pedestrianism came from the introduction of new pavements to speed people on their way and cigarette machines and the brightly-lit milk bar encouraged progress.[37] The London underground's newly installed 'ticket-and-change machines' and escalators encouraged the business of getting to work. Sybil Andrews's linocut print called *Rush Hour* (1930) idealized well-toned purposeful strides as they met the sleek surfaces of the regimented commuter route (Figure 4.1). So much of the sensory experience of life in London and other cities was ruled by the passage through the interiors of home, public transport, workplace, or the shops. Though the dancehall attracted many young working-class and lower-middle-class men and women to free themselves to the sound of American-style jazz and British dance music at the weekend, expression was largely prescribed and it was still contained indoors.[38]

An unregulated freedom of flow was found in rambling. Legs were no longer simply for transport but propelled the body in pleasurable exploration. Walking in nature brought an 'equable and abundant flow of tranquil and half-conscious meditation', argued Leslie Stephen.[39] This mental and physical state was hard to find elsewhere. For the historian G. M. Trevelyan, modern life with its machinery

Figure 4.1 *Rush Hour* by Sybil Andrews (1930).
(© Glenbow, Calgary, Canada. © The Trustees of the British Museum)

had denied the human race 'the natural use of its limbs'.[40] One's own rhythm could not be known in the city because there were simply too many competing and conflicting rhythms of other humans and machines. In the countryside, however, with the arms swinging, the rhythm of body and mind would present itself. To attain 'the harmony of the body, mind, and soul', Trevelyan preferred to be alone, finding his own natural pace.[41] In this condition perception opened up.

The humble footpath gave the most 'intimate acquaintance with the country' and the closest relationship to the soil said the Birmingham-based rambler Edgar Brooks. For him, the best paths were the ones that were hardest to spot. He would look out for where 'the narrow brown streak grows fainter, until the only sign of a path is a slight discolouration of the grass'. To plunge down an unknown path was a thrill. Left far behind was the sensation of foot interfacing with pavement or tarmacadam. Now the boot could feel the spring and cushioning of grass, or it could leave an impression in the mud.[42] Rambling on the road was not out of the question and Trevelyan favoured these surfaces for their 'invaluable pace and swing', which helped one to cover long distances. He enjoyed 'the change in tactile values under foot': the 'hard road', the 'unfenced moorland road', the 'common farm track', and the 'broad grass lanes of the low country'. All had their rewards and the mix was enchanting.[43] For a serious walker like Trevelyan though the true seductions of nature could only be found in a cross-country adventure. To 'scramble', however, was to abandon the rules and techniques of walking altogether in order to find the greatest sensory pleasures:

> Scrambling is an integral part of Walking, when the high ground is kept all day in a mountain region. To know and love the texture of rocks we should cling to them; and when mountain ash or holly, or even the gnarled heather root, has helped us at a pinch, we are thence-forth on terms of affection with all their kind. No one knows how sun and water can make a steep bank of moss smell all ambrosia till he has dug foot, fingers, and face into it in earnest.[44]

The scramble allowed the sensuousness of nature to be properly encountered, while the body pulsed and sweated in ways it could not in the town or city. The rambler's body was foregrounded in consciousness, because it did things and felt things not usually experienced at home or at work. The daily chatter of the mind might settle or be pushed aside by other impulses.

A dimming of self-consciousness allowed the rambler to lose themselves in the tangle of the natural world. The Londoner Minna Keal revolted against the conventionality and boredom of domestic life and 'revelled in the freedom of unruly open space'. Energy, drained by urban routine, was replenished. There was, she found, delight in getting caught up in mountain mists or lost in the darkness of the New Forest.[45] As the Clarion Ramblers' motto ran: 'The man who was never lost never went very far'. The spread of way-marking was seen by some ramblers as an unwanted domestication of nature and a further intrusion of modern ways. It was the contours of the landscape and its features that best marked out wild nature.[46] Some map enthusiasts who wanted to know where they were going were able to see in the contours and markings on the flatness of paper, not just the reassuring rationality required for navigation, but also something sexy. The journalist and mountaineer CE Montague had this to say about the aliveness of the map in his 1924 essay 'When the Map is in Tune':

> The notation once learnt, the map conveys its own import with an immediateness and vivacity comparable with those of the score or the poem. Convexities and concavities of ground, the bluff, the defile, the long mounting bulge of a grassy ridge, the snuggling hollow within a mountain shaped like a horseshoe—all come directly into your presence and offer you the spectacle of their high or low relief with a vivid sensuous sharpness.[47]

Rambling allowed people to feel again. And the sensation of quiet in a country walk was not always what was sought out. Minna Keal had too much quiet at home and wanted the tumult of the untamed in her excursions—'the thunder of the waterfalls, the turmoil of the mountain cataracts, the melancholy echo of the sea'.[48] To have the senses disrupted in bad weather brought unexpected but welcomed impressions too. For instance, the loss of sight in mountain mist brought the pleasure of listening to the fore Trevelyan observed: 'Listen while you can to the roar of waters from behind the great grey curtain, and look at the torrent at your feet tumbling the rocks down gully and glen, for there will be no such sights and sounds when the mists are withdrawn into their lairs [. . .]'[49] Carrying on in rough weather may have been a point of honour, a mark of 'hiking heroism' especially north and west of the Midlands, but it was also a thrill which set the skin tingling.[50] Claude Fisher captures the theme in his *Daily Mail* rambling column: 'feel the wind and rain beating on one's face and one's body aglow with health and energy'. Fisher advises readers to add extra underclothing but to 'stick to shorts', which kept bare skin exposed to the elements.[51]

From all this it is apparent that human senses were still functional in the towns and cities of Britain even if they were operating defensively at times. The reports of the challenges of urban life make clear that the senses were by no means switched off. Indeed, they were readily revived when exposed to the natural world. Next is an analysis of the highly personal accounts of two ramblers. Both accounts stem from long immersion in nature on foot and are of course informed by prevailing social and cultural conditions. They provide fresh perspectives first from a young man newly exploring the Derbyshire moors and second a woman in her forties reflecting on decades of walking at high level in the Scottish Cairngorms. He was a silversmith, she a teacher and novelist, and both report from the regions of Britain most devoted to rambling.

Willis Marshall: Into the moors

We all forsook our duties to our own bodies when we accepted what we call civilisation, we had our natural beauties exhibited in a stuffy cinema... but how many took the trouble to go and seek them for themselves.[52]

George Willis Marshall started rambling when he was nineteen, going out into the Derbyshire Peak District with three or four friends from 1923 (Figure 4.2). They lived in Sheffield so naturally enough joined G. H. B. Ward's socialist Clarion Ramblers club, continuing to walk with the club and taking part in the Kinder Scout trespass. From his first days of serious rambling in the early

Figure 4.2 Stopping for lunch at Jaggers Clough on a twenty-five-mile walk on Sunday 28 October 1923. Willis Marshall is second from the left.
(Courtesy of Willis Marshall's family)

1920s Marshall kept two carefully written walking journals, which are examined here.[53] They are full of the excitement and camaraderie of young men exploring their world together and are written at a time that preceded the popularization and demystification of the land that came with 1930s rambling movement. Each of the ninety-five walks is given a page or two in which he recorded the route and distance, start and end times, where 'dinner' and 'tea' were taken, and dotted throughout are detailed pencil drawings of an isolated pub or ancient standing stone, along with his own photographs pasted in. There are notes about meeting a group of 'tramps' along the way and spotting the flash of a kingfisher for which they 'felt highly honoured'. Marshall's main journal was called 'Rambles from Sheffield into Derbyshire, January 21st 1923 to September 28th 1924'. Ward's resonant words from the *Sheffield Clarion Ramblers Handbook* are written out on the title page: 'A rambler made is a man improved' and 'The man who was never lost never went very far'. And there is a poem: 'Free is the bird in the air / And the fish where the river flows / Free is the deer in the woods / And the gypsy wherever he goes / Hurrah! / And the gipsy wherever he goes'. Marshall's craftsmanship as a silversmith is evident in his meticulous and artful journal making.

The hill walks were long and hard, often starting at 8am, and the return was in the late evening, sometimes after dark. These were not the Sunday rambles that Londoners were taking in the Home Counties; rather they were characteristic of the Northern appetite for strenuous walking at height. The 'forbidden' northern moorland around Kinder Scout where Marshall would trespass is a wild plateau of about fifteen square miles, gouged by deep peat trenches. Distances of twenty-five miles were typical. A fourteen-mile trip was described as a 'lazy day'. Sometimes they took a train to the starting point, on other occasions they would walk from home in Sheffield. Like the Clarion handbooks of Ward, Marshall's journals are full of inspirational Romantic poems from Blake, Shelley, and Crabbe, but also from local friends and ramblers. Most are about the experience of the natural world—the tiny bell-shaped heather flower, autumn colours, the excitement of the hills, the waters of the river Dove. Faithfully copied out, these favourite poems speak for his interest in the sensations of nature. Having scrimped and saved, the friends set out for a week on a 'walking tour' of 120 miles in August 1923. Stan Oxley, who was sixteen, wrote the journal this time with Marshall adding sketches and poems. The rhythms of the poems complemented the rhythms of the rambling trip: Marshall writes out a verse from Tennyson's 'The Brook' which starts 'I chatter, chatter, as I flow'.[54] The weather is given a mnemonic verse with its own rhythm: 'Dirty days hath September / April, June and November / All the rest have thirty one / Without a single gleam of sun!'[55] To keep the pace going and to lift the mood, Marshall and his friends sang popular songs of the day. The jaunty American foxtrot 'Oh! My Sweet Hortense' was a favourite, Stan accompanying on the tin whistle.[56] It was a flowing song in 4/4 time that they would have heard on the gramophone. The chorus went like this: 'Oh! oh! oh! My sweet Hortense / She ain't good lookin' but she's got a good sense / Before I kiss Hortense / I always buy a nickels worth of peppermints.'

On the week-long trip they over-nighted in last minute digs, under haystacks, in barns, sometimes under the stars in the heather. This nocturnal exposure brought a submersion in nature that felt profound. But it was full of strange encounters too: burying himself in the hay one night, Marshall's friend Oxley reported falling asleep listening to 'my little bed mate, a young rat hunting for food'.[57] The summer after in June 1924 the group did a 'midnight ramble' that started at 10pm on Saturday and finished on Sunday evening, having walked forty-five miles. It was a 'brilliant moonlit night' Marshall wrote in his journal.[58] Night walks became popular, even fashionable, in these years. The writer S. P. B. Mais led six hundred London hikers on a guided Harvest Moon ramble to Ditchling Beacon on the Sussex Downs in early September 1932. It was not a success however, with no moon and no sunrise. The newspapers enjoyed reporting that 'the moon had set before the train reached Hassocks and the sun did not appear until after lunch'.[59] The darkness gave a chance to listen though. Claude Fisher in the *Daily Mail* wrote of the pleasures that only the rambler who had escaped from the city could have: ' "freeze", and you will have a rich reward [. . .] only then, when you are as still as a rock, will you know that every hedge and ditch, every tree and field teems with life, but little if ever revealed to the human slaves of routine'.[60] These guided excursions in the night were attractive to courting couples who were 'walking out' and receptive to the intimate potential of darkness in the gentle landscape of southern England (Figure 4.3).[61]

Willis Marshall and his friends went further to get in touch with Derbyshire's natural environment. Like other ramblers, they had an affinity for the water of the pools, falls, and brooks that snaked through the moorland. They were drawn as well into the caverns that fissured the landscape with thrilling and dangerous openings. Dove Dale was the most beautiful valley Marshall and Oxley had ever seen. Marshall copied into the journal Charles Cotton's poem from the Complete Angler about the famous trout stream called the Dove.[62]

> Oh my beloved nymph fair Dove
> Princess of rivers how I love
> Upon thy flowery banks to be
> And view they silver stream
> When gilded by the summer's beam

" EXCUSE ME, BUT HAVE YOU SEEN ANYTHING OF A CONDUCTED MOONLIGHT RAMBLING-PARTY ? "

Figure 4.3 A cartoon from *Punch* magazine, August 1936, with the caption: 'Excuse Me, but Have You Seen Anything of a Conducted Moonlight Rambling-Party?'.

To write out and memorize such poems was to express sentiments that otherwise might have remained unsaid. They were part of Ward's mix of progressive self-discipline and athleticism that co-existed alongside a Romantic and emotional turn of mind that gloried in the freedom of the wild moors, that fought against the griminess of town-life and workplace burdens.[63] In Rudyard Lake the friends took a refreshing dip, having asked permission at the angling club hut: 'Bill was hot and desirous for a swim, so he asked the fisherman if he may bathe in the lake without offending anyone.'[64] The day before, Stan had fallen in a stream trying to catch a crayfish for breakfast. There is a boyish but masculine tenor to these adventures that is found in other accounts of walking in the Edwardian and post-war period. G. M. Trevelyan wrote of the 'northern torrent of molten peak-hag that we must ford up to the waist' and the 'plunge in the pool below the waterfall' as the essence of properly experiencing nature's tactile beauty.[65] Ward, a mentor to Marshall and his rambling companions, was a keen bather and allowed himself to be photographed revealingly (Figure 4.4). The first Clarion ramble he led in 1900 was recalled by a member of the group: 'Now the descent of Ashop Clough was before us, and after persuading the ladies to go on ahead, several of the men indulged in a bathe in one of the deeper pools'.[66] It is clear that naked bathing was part of Ward's masculine Clarion regime. His Clarion handbooks, which Marshall and friends knew well, often dwelt on the physical cleansing and spiritual refreshment to be found in the 'virgin' purity of moorland streams and waterfalls.[67] Dave Sissons has suggested that Ward's interest in bathing in the nude came in part from his friendship with and the socialism of Edward Carpenter, who introduced Ward to Walt Whitman's controversial poetic sensuality. Carpenter, who lived in Millthorpe in the Cordwell Valley, south of Sheffield, liked to walk naked in the woods there 'acting out his philosophy of freedom and naturalism'.[68]

These bodily practices of physical exposure to the elemental forces of water, air, and sun are examples of sensual engagement with the natural world that was a key part of the rambling scene, and not only in the vicinity of Sheffield. In C. E. M. Joad's *The Charter for Ramblers* he endorsed the physical exposure of the body to nature, along with the mind, in the 'making of whole men and women'. Rambling in nature was the best way to educate the whole person. 'The feeling of the air upon the skin, of the sun upon the face; the tautening of the muscles as we climb; rough weather to give us strength, blue skies and golden sunny hours to humanise us – these things have their influence on every side of our being.'[69] This was not the socialist faddism or perversion that some like Orwell charged characters including Carpenter with ('fruit-juice drinker, nudist, sandal-wearer, sex-maniac, Quaker, "Nature Cure" quack, pacifist and feminist').[70] This was disciplined, progressive, moral enlightenment.

Figure 4.4 G. H. B. Ward taking a dip in a pool near Swains Greave on the Bleaklow moorland, Derbyshire.
(Courtesy of South Yorkshire and North East Derbyshire Area Ramblers. Photographer: Harry Diver c. 1910–20)

Marshall with several others found a way to get yet closer to the moorland they loved by squeezing into its underground caverns and passages. The bedrock of this part of England was fractured with strange inner workings that the Yorkshire Rambler's Club had been venturing into since the turn of the century.[71] Gaping Gill and Alum Pot were notorious black holes in the limestone that Marshall would have read about. Two particular caves drew in Marshall and his comrades: the Wonder Cave and the Peak Cavern. His journal records a four-hour exploration of the Wonder Cave during Ramble 75 on Sunday 4th May 1924 and includes a set of sketches of the layout of the cavern system with approximate dimensions and ceiling heights, the result of several underground trips. In the following years they explored Old Tor Mine near Winnats Pass and Swallet Holes at Rushup Edge. Communion with the innards of the hills brought 'much grunting and groaning, howls of anguish and bodily pain—also a few pointed remarks about the hardness of the rock'.[72] Sometimes Marshall collected rock samples and took them home. His photographs show the state of the over-alled cavers when they came to the surface, scuffed and muddy but victorious.

Wrestling with the rocks was to reckon with a wild nature that was, in Ward's mind and therefore others, masculine. In turn, stoicism and hardiness, which were central to ideals of manliness, were needed to assert oneself in these sur-roundings.[73] This did not exclude the expression of emotional or poetic feelings—the realization of nature as a deeply sensuous experience required both a romantic sense of being overwhelmed hand-in-hand with the security of manly domination. On their rambling trips, Marshall and Oxley wrote how they 'set out as though it was darkest Africa we had to conquer', seeming to reflect a con-temporary imperial and masculine discourse of exploring the unknown and the fantasy of escaping the tameness of the everyday.[74] Yet the heroism is tempered with sensitivity in contemplating their return from a week on tramp when Oxley bemoaned in lyrical lines that 'the stately pine caressed by a perfumed breeze would give way to gas lamps and tram standards. The call of the peewit would be replaced by the six o'clock buzzer whose raucous voice oft called me from my bed'.[75] The escape had been full of wonderment but it was over for the time being. The sensory impressions of the natural world would have to be exchanged for the more familiar ones of working life in Sheffield.

Nan Shepherd's merger with the mountain

G. H. B. Ward did not want to see women on the hardest of his Clarion Ramblers walks, writing about the Sunday Revellers' Ramble at Cut Gate in January 1926: 'We go wet or fine, snow or blow, and none but the bravest and best must attempt this walk. Ladies, on this occasion, are requested not to attend'.[76] Ward was ambivalent about women's rising status and visibility in society, but women were tackling mountains as well as hills with their husbands, and increasingly in their own organized groups. The Ladies' Alpine Club was formed in 1907 and a year later the Ladies' Scottish Climbing Club.

Nan Shepherd was not one for clubs, yet she spent most of her life climbing in Scotland's wildest range, the Cairngorms, which has five of the six highest peaks in Britain. Shepherd climbed all six summits, some twice over, but began to find the elation of the peaks exceeded by the sustained height of the Cairngorm pla-teau ('the mountain', she called it).[77] 'Summer on the high plateau may be as de-lectable as honey; it can also be a roaring scourge', she wrote of the elemental extremes of this place.[78] Both the honey and the scourge were good, however, be-cause each was part of the essence of the mountain world. Shepherd's impressions of being on and with the mountain for almost forty years are vividly set out in her short non-fiction work *The Living Mountain*, which she wrote during the period of WWII. Her profound experiences of the natural world of the Cairngorms fed her writing life—she was a novelist and English teacher at Aberdeen's teacher training college—and *The Living Mountain* is her attempt to bring into words the felt experience of the landscape that she knew so well. As such it is a document of historical note for its explication of how nature might be sensed and made sense of in the 1920s and 30s.[79] At first the book was to be called *The Plateau* but her most trusted author friend, Neil Gunn, read the manuscript and said it sounded 'a trifle square-lined [. . .] you are dealing with curves and penetration'.[80] Indeed, it is a startlingly personal account of the Cairngorm environment perceived by a woman who was able to achieve a state in which senses, mind, and body combined with the mountain environment.

In those mountains, she wrote, 'may be lived a life of the senses so pure, so untouched by any mode of apprehension but their own, that the body may be said to think. Each sense heightened to its most exquisite awareness is in itself total experience'.[81] With her thinking body, all senses switched on, Shepherd was able over the years to make contact with what she called the 'total mountain', which included its geology, plants, and animals—'the grumbling grinding mass of plutonic rock' overlain by scree, soil, water, moss, grass, trees, insects, birds, and beasts. Shepherd wondered, as if making a philosophical enquiry into the nature of knowledge, what else she might discover if she had additional senses: 'There must be many exciting properties of matter that we cannot know because we have no way to know them'.[82]

The process of walking was required to meld the senses with the mountain; walking steadily for hours, 'with the long rhythm of motion sustained until motion is felt, not merely known by the brain'.[83] The rhythm of the legs combined with the rhythm of breathing until Shepherd was able to reach a 'profound harmony' that could deepen into something like a trance:

> Walking thus, hour after hour, the senses keyed, one walks the flesh transparent. But no metaphor, *transparent*, or *light as air*, is adequate. The body is not made negligible, but paramount. Flesh is not annihilated but fulfilled. One is not bodiless, but essential body [. . .] I have walked out of the body and into the mountain.[84]

This is more than an immersion in nature that Marshall and Oxley, and Ward and Trevelyan had experienced. Shepherd experienced a kind of meta-physical merger. Senses had connected together, the divisions between mind and body had dissolved and Shepherd's 'Being' flowed into the natural world of the Cairngorm range. There is a kind of mysticism in Shepherd's writing informed as it is by her reading of Eastern and especially Buddhist thought. Spiritual and sensuous experience co-exist, and yet Shepherd's accounts of nature are grounded. For instance, when Shepherd first climbs Ben MacDhui, the highest peak in the range, she is astonished to find the mountain can have an inside. At the end of the climb via the Coire Etchachan route, instead of spaciousness and vista, she finds an interior. Loch Avon is below, 'and on every side except where we had entered, towering mountain walls' with the 'soaring barricade of Cairn Gorm beyond'.[85] On another occasion, Shepherd has a far more profound encounter with the interior of the mountain when looking into the clear water of the Loch Avon that rests at 2,300 feet. With her unnamed companion she undresses. It is a cloudless day in early July.

> The clear water was at our knees, then at our thighs. How clear it was only this walking in could reveal [. . .] We waded on into the brightness, and the width of the water increased [. . .] Then I looked down; and at my feet there opened a gulf of brightness so profound that the mind stopped. We were standing on the edge of a shelf that ran some yards into the loch before plunging down to the pit that is the true bottom. And through that inordinate clearness we saw the depth of the pit.[86]

Shepherd had seen inside the unknowable body of the mountain. These sensations are of course not devoid of fear, and nor were those of Willis and his mates when caving underground. Perhaps the closeness to danger was indicative of the commitment being made to forge a relationship with a complex natural environment. In Shepherd's case, this relationship extended to conversation with the natural world. In morning silences Shepherd finds 'water is speaking'.[87] If listened to, water made music too: 'the slow slap of a loch, the high clear trill of a rivulet, the roar of a spate. On one short stretch of burn the ear may distinguish a dozen different notes at once'.[88] The ever-changing ancient songs of cold water over stone was the subject of her 1933 poem called 'Singing Burn'. Such sounds penetrated into Shepherd's consciousness and inevitably she spoke back to them, for she remained mystified by the properties of mountain waters and wanted to understand them.

For Shepherd, any notion of merger or becoming one with mountain life was based on the absorption of sensations, which sometimes might result in an almost erotic fervour. Shepherd's first interests as a walker were 'only sensuous gratification—the sensation of height, the sensation of movement, the sensation of speed, the sensation of distance, the sensation of effort, the sensation of ease: the lust of the flesh, the lust of the eyes, the pride of life'.[89] She says 'only', because later that selfish appetite for the sensuous became deepened as she began to know and understand the 'total mountain', its contours and colours and life.

As a young woman she wrote: 'I was not interested in the mountain for itself, but for its effect upon me, as puss caresses not the man but herself against the man's trouser leg'.[90] Shepherd put herself in the way of mountain stimuli, getting into the granite cold water of a stream, standing until the sensation left her legs. Or she walked bare foot across scratchy heathland or after rain ran her hand through juniper for the joy of the wet rain drops trickling over the skin. 'Touch is the most intimate sense of all', she declared.[91]

There is frank animality in Shepherd's passions. 'I am like a dog—smells excite me', she writes. Soil, the activity of bacteria, the drip of fir tree sap yield smells that signal 'the business of living'.[92] Shepherd relished the earthy smell of moss, 'best savoured by grubbing' as well as the scent of heather rose surrounding her.[93] The high cries and mighty curves of a swift in mad flight on top of the plateau 'invades the blood' to the extent that she feels she shares in the bird's motion.[94] More fundamental than animality there was recognized by Shepherd a life force in the natural system she loved. She saw the surges of water in the rocks 'as integral to the mountain as pollen [is] to the flower', an evocation of the energetic potency of water and its role in shaping the mountain's future life.[95] G. H. B. Ward had imagined similar masculine urges at work in the waters of Derbyshire's Kinder Scout, Melanie Tebbutt has pointed out, 'Kyndyr' according to Ward meaning 'head of the waters' and 'Scout' referring 'to pouring forth a liquid forcibly'.[96]

In the early part of the twentieth century, few women wrote like this about their experiences of nature.[97] Robert Macfarlane has noted that Shepherd articulated her ideas of the sensing, thinking body and embodied knowledge at the same time as Maurice Merleau-Ponty was publishing *The Phenomenology of Perception* in 1945, yet without the elite institutional support that surrounded the French academic.[98] Shepherd does not write like Merleau-Ponty, however, and when prose is as beautifully composed as Shepherd's is, there is a danger of losing critical distance. There is also a danger of assigning to her writing a particularly female sensitivity, although we have seen here that male walkers and writers could think and feel in similar ways.[99] William Wordsworth's expressiveness was emblematic of a tradition of men writing about nature with feeling. Shepherd would have been influenced not only by his work, teaching it to her college students in all likelihood, but also Wordsworth's rambling practices.[100] Shepherd's writing is distinct from nineteenth-century transcendental styles that drew upon individuality and freedom but also the divine, which Shepherd does not invoke. While there are Zennish interpretations in some of her encounters, there is at the same time reasoned thought to match. Others who write about walking in the 1930s do think about nature in explicitly mystical terms. Frederick Wills, for example, a columnist for the *Newcastle Evening Chronicle*, wrote of 'the revelations of the infinite spirit', believing in 'worlds unknown in the countryside' and unseen earthly vibrations emanating from ancient rocks and earth.[101]

Shepherd usually walked alone, sometimes seeing no one all day—'I have heard no living sound [. . .] up here, no movement, no voice. Man might be a thousand years away'.[102] The intensity of her relationship with the mountain must have come from such isolation. *The Living Mountain*, then, is highly personal account yet it is not detached from the experiences others had. Many other walkers were drawn to the hills and mountains around Aberdeen who would have had their own intense encounters with the natural world. Shepherd knew some of these people herself. 'There are addicts of all classes', she wrote, referring to the mountain walking bug. The hill lovers who really know the contours and dangers are 'shopmen and railway clerks and guards and sawmillers'. On one occasion she meets two lean nineteen-year-old railway workers who have come up from Manchester to spend their one week's leave to try to photograph the Golden Eagle. There are others like a bony-kneed man in kilt and Highland cloak, a 'red-headed greaser, an old mole-catcher, and an errand boy from Glasgow'.[103] The kinds of sensory and bodily experience that Shepherd wrote about were available to all people, and so we can take Shepherd's account of the natural world as one that reflects what many might have confronted when rambling in interwar Britain. She does not speak for others, but her writing brings to light feelings about nature that were not uncommon. Her writing makes clear how interconnected sensory experiences can be—listening to nature can sometimes hardly be teased apart from the other inputs to perception and may make less sense when we try to.

A violent assertion of personality: Hedonism in nature

The character of the outdoors, its natural settings and peaceful acoustics, could be a stage for celebration and release, a trigger for talking and singing and other sensory indulgences. For many modern British men and women, the natural world was an environment in which they could assert themselves in ways they could not at home or at work. While nature was not worshipped by everyone who went there at the weekend, it was not ignored either. In fact, being in nature had a powerful effect on many rambling urbanites according to C. E. M. Joad: 'the presence of wild Nature prompts them to reassure themselves with violent assertion of their personalities'.[104] Though Joad was talking about 'day-trippers' or 'amateur ramblers', who he saw as vulgar and careless and in need of education, exuberant expression could overcome anyone confronted by the country in the raw. Nan Shepherd found that her first encounter as a teenager climbing alone in the 'forbidden country' of the Cairngorms released her inhibitions: 'I gulped the frosty air—I could not contain myself, I jumped up and down, I laughed and shouted.'[105] In his philosophy of walking Frédéric Gros has pointed out recently how walking in nature can provoke excesses of emotion or a sense of drunkenness, noting that a surfacing of animal presence aroused in the outdoors can bring forth a shout from inside the body.[106] The middle-class portrayal of ranks of interwar ramblers as disengaged from their natural surroundings can be rethought by arguing instead that these people brought nature into their lives on their own terms that by no means excluded its sensory pleasures.

Accounts from revellers themselves are obscured by the deluge of consternation from those who questioned the quality of people invading the countryside. The newspapers were busy with debate about the kind of people that constituted 'true ramblers'. They were not the ones who played leap-frog on the railway platform, wrote a Barnsley rambler. There was nothing wrong with boisterous energy but it was better dissipated on a fifteen- or twenty-mile tramp over rough moorland.[107] According to a columnist writing in the *Manchester Guardian* in May 1933, some ramblers wanted 'party solitude, well provisioned and equipped, with some noise-making apparatus', which was usually a wind-up gramophone player.[108] Party solitude was presumably an atmosphere in which you could gather with your friends in high spirits without being disturbed or told off. Common-place contemporary attitudes to the behaviour of the wrong sort of rambler are revealed in this poem that appeared in the 1929 handbook of the London Federation of Rambling Clubs:[109]

> A Brace of Trippers underneath a Bough
> Some Orange Peel, some Gasper Cards and Thou
> Broadcasting Bellows straight from 2LO
> Make every glade a little Hell now.
> Think from each reeking charabancserai
> The whole of Summer, shrieking night and day
> How Tripper after Tripper, with his Girl
> Will litter every lane and go his way.

Through the tone of condescension, picnicking, singing, and full-blooded companionship are in evidence here, as is the music from a portable wireless set broadcasting the BBC's 2LO station. These are the precious pleasures of people taking time off work, away from the burden of daily responsibilities, pleasures that can be afforded only once in a while. That so much writing of the time dismisses the behaviour of the lower and lower-middle classes as crude and in need of reform can disguise the many social meanings at play.[110] The third-party accounts of rambling day-trippers in fact demonstrate the free-wheeling sensory celebration of modern Britons making use of the natural world, albeit in sometimes irreverent ways. To bring the 'atmosphere of the town into the country' as

Nora Willington, General Secretary of the Manchester and District Ramblers' Federation scathingly put it, was part of the point.[111]

The fun would often start on the train, or in an open-top bus called a charabanc and a country pub would inevitably be visited during the day.[112] Train and bus companies expanded their timetables to accommodate ramblers and aimed their advertising at urban people who were out for a good time in the countryside. Many advertising graphics showed a man and women in shorts and bare arms striding out with purpose. But some acknowledged the potential for rural hedonism. The WWI war artist C. R. W. Nevinson showed a group of well-turned-out men and women bounding along hill and vale singing and dancing to music played on a concertina and mouth organ (Figure 4.5). Nature is key to this scene of pleasure—it is not simply an attractive backdrop, it has a role in activating the spirits with its textures, sounds, and fresh air.

Taking pleasure in the rambunctiousness that came with larger walking groups was not beyond those who rambled within the officialdom of their clubs. South Shields MP James Chuter Ede told of the gregariousness of open-air fellowship in 1938: 'I frequently go out on the Sunday rambles organised by the Southern Railway and I know of nothing more exhilarating even when there are 700 or 800 people on these rambles'.[113] Because the freedom of the hills brought the body and mind so passionately to life, clubs and national organizations, like the Ramblers' Association and Youth Hotels Association, devised their own codes of conduct to help ramblers manage their excitement and to preserve the beloved fabric and mood of the countryside.[114]

The tradition of not talking while walking, favoured by earlier generations, was no longer prevalent. Robert Louis Stevenson in his essay *Walking Tours* wrote strictly that, 'there should be no cackle of voices at your elbow, to jar on the meditative silence of the morning'.[115] The seasoned walker Stephen Graham writing in 1926 on the other hand found that rambling rhythms encouraged 'naturally long conversations', which meant one got to know a walking partner very well. 'You comment on Nature around you, and on tramping experiences'.[116] Increasingly, rambling brought a song to the lips, even of the lone wanderer. Unlike talking, another party was not needed. Singing out loud was a privilege of being alone in the countryside where there were no critical ears. Stephen Graham worried that though 'man is a singing animal, but civilisation has silenced many songs'. He pointed out how annoying it was if someone began to sing on the train. Urbanites like him craved 'silence and our newspapers, silence and our pipes'.[117] For Graham, the singing impulse was a very natural thing when he took to the road. The stamping of the feet provoked with its rhythm the emergence of a song and equally the feet would synchronize once a song was in train.[118] Any song would serve the walk, even a tune from childhood or schooldays, or a jumble of half-remembered melodies.[119] Graham liked the folk song 'The Girl I Left Behind Me',

Figure 4.5 An advertising poster for bus travel illustrated by C. R. W. Nevinson in 1921 with the caption: 'Pleasure. One Way to Pleasure – By Motor-Bus'.
(© TfL from the London Transport Museum collection)

an old military lilt about leaving behind 'moor and valley'. It is a tune he may well have sung during his time in the army.[120] Semi-official songs like Ivor Novello's 'Keep the Home Fires Burning' or 'Goodbye-ee' were sung by ex-servicemen and civilians alike. These kinds of songs and improvised ones from the march, barracks, and trenches were gathered together in bowdlerized versions in the *Daily Express Community Song Book No. 3* called 'Songs that Won the War' in 1930.[121] The songs that sustained soldiers and civilians during wartime still had legs over a decade later and are certain to have been aired on rambles.

Plenty of popular songs from the metropolis would have been taken out into the country. Jazz, swing, and big-band dance were the new sounds on the radio and circulating on gramophone disc—Joad had suffered the 'disjointed rhythms of some American negro song' coming from a loudspeaker on one of his trips.[122] The potent tradition of music hall song was still in the blood though, especially of the working classes, and that style gave rise to many new stars including George Formby, Gracie Fields, Max Miller, Will Hay, and Flanagan and Allen during the 1930 and 40s.[123] Music hall strongly influenced the vernacular songs made by factory workers, cotton spinners and weavers and coal miners out of their own experience and passed around orally.[124] What should be noted here is that the countryside, with its age-old soundscape, would have also resounded with the modern sounds of songs brought from the human world of dancing, working, and fighting. And part of the sonic culture of walking in nature was as much about marking out a rhythm or a tune as it was listening to a bird in song.

Ramblers made their own songs about rambling of course (especially in the north of England). Ewan MacColl's 'Manchester Rambler' became an iconic song in the early 1930s. It was a political song, inspired by his participation in the organized mass trespass onto the Kinder Scout peak in Derbyshire in 1932, a protest of hundreds by the urban Young Communist League of Manchester. With more widespread appeal was the upbeat 'I'm Happy When I'm Hiking', a communal song celebrating nature and popularized by the *Daily Herald's* League of Hikers.[125] Clubs published their own song books as did local scout groups.[126] These songs often referenced the natural world and its beauty rather than detract from it.

Stephen Graham was clear how the soundings of a day spent in nature would stay with the walker and might spark the desire to sing in camp later on. Once the campfire was lit, he wrote, 'you are full of the songs of the birds to which you have listened all day. Music will come out again.'[127] What can be taken from this is that the singing of songs on foot did not necessarily obscure the sounds of nature. There could be a kinship between making sounds in nature, feeling one was part of nature and being attentive to its sounds. Having said this, though nature was a place of escape it was not necessarily for its quietude. Many people wanted to sing and play and make music. Yet, there was still sensory contact with nature as one walked and picnicked in it. The often-criticized picking of wildflowers—bluebells or primroses or cowslips—was a way to grasp the feel and aroma of the natural world, a sensuousness that could be taken back home to remember a day out in the countryside.

Conclusion

By taking an approach here that considered all bodily sensations and their interplay, it has been possible to find much more than a mood of peace and quiet in rural Britain. It has been possible as well to expose some of the fundamental sensory urges that propelled trespassers, day-trippers, and rural preservationists to action, and united them, even though these groups were often at loggerheads about how best to enjoy nature. It was the sensuous experience of nature that was sought out by these groups and this experience was produced through the proximate sense of touch, in particular, as well as hearing. This approach has also made clear that not everyone listened to nature during this period. Why should they have? There were many ways to engage with one's natural surroundings and attention was not always directed towards listening. In fact, the natural world could itself provoke the rambler into making their own sounds, to complement walking rhythms or to celebrate a sense of freedom, and this contribution to the soundscape should be considered as part of a wide-ranging sensory interplay of body and mind with outdoor surroundings. Listening cannot be properly understood without consideration of the other senses operating at the same time that create together what we call the *feeling* of being in nature.

The popularity of interwar rambling was indicative of a powerful interest in exposing the senses to the many varieties of British nature. In particular, challenging upland terrain presented opportunities to put the body to work amid the rock and water of the landscape, with unpredictable weather bringing further delights of sensation to the body. Almost everyone agreed that 'escape' from urban and industrial environments was the main motivation to take to the hills, but this was not simply to relieve over-burdened senses. Equally, walking in nature brought to domestic and work-bound lives the alternative sensations and stimuli of extraordinary vistas, the rush of mountain water, or a tingle to the skin in contact with heather or cold rain. Some ramblers wanted to do more than experience the feelings of being on the surface of the land: getting into and inside it was perhaps the most thrillingly intimate.

More than an escape from or rejection of modernity, rambling in British nature was a means by which men and women could have a controlling influence upon the sensory repertoire in their lives. It was the modern travel modes of train and bus that were instrumental in establishing weekend rhythms of exchange between urban and countryside experiences. Reflecting on his survey of England in 1933 by motor coach and car, J. B. Priestley imagined a future that combined the beauty and tranquillity of 'Old England', the vigour and independence of the 'England of the Industrial Revolution' and the classless, democratic spirit and modernity of 'Postwar England'. In other words, Priestley hoped for a progressive balancing of old and new, an idea which was often exemplified by the rambler's modern mentality.

In the late 1930s the government's National Fitness Campaign called on all citizens to participate in sport, gymnastics, and outdoor recreation including 'hiking' and cycling. Hiking, rather than the more sedate-sounding rambling, fulfilled individual needs but also political ones as it was seen to be a useful preparation of citizen bodies and minds for war.[128] Hiking could do more than improve fitness for military service—it got citizens used to being outdoors where the campaigns of warfare were conducted, and it brought knowledge and understanding of the landscape that was soon likely to be under threat.

5

Modern birdsong and civilization at war

> Even if we be mechanizing so many aspects of life, or rather, just because we *are* mechanizing them, there is all the more reason to reserve to birds—shy birds as well as tame, rare birds as well as common—a place in our civilized scheme of things.
>
> Julian Huxley, *Bird-Watching and Bird Behaviour* (1930)[1]

In the mid-1930s, technological advances in sound recording meant that for the first time the vibrations of wild nature could be apprehended and then fixed onto a ten-inch gramophone disc in high fidelity. The process of recording took the elusive moments of birdsong and solidified them so that the owner could indulge in repeated listenings. But what was lost when the ephemeral notes of nature were captured and committed to a medium for consumption in one's own time and place? And to what cultural and social uses were such recordings put in peacetime and in the crisis of another war? First, I consider the contribution of the key exponent of wildlife sound recording, Ludwig Koch, who was responsible for recording and modernizing British birdsong with the help of Julian Huxley and Max Nicholson. His work allowed new kinds of concentrated listening and knowledge-making to take place by a wide public as well as by ornithologists. Second, I evaluate the functions and reception of these representations of birdsong when they were put to work on BBC radio during wartime, and so became a state tool for managing the feelings of citizens at home. The goal is to assess the wider question of the place of bird voices in British culture when the certainties of a civilized world under the care of human voices are breaking down. Many quarters of society in the late 1930s suffered from what has been described as 'war psychosis', a state of mind in which the question of the survival or death of contemporary civilization was at stake.[2]

Throughout WWII, BBC broadcasting sent Ludwig Koch's birdsong into millions of homes at a time when moods were more febrile still and public listening attention was newly geared towards keeping safe from attack from the air, especially in the cities. In wartime, the radio was a vital source of news about some of the realities of war, as well as relief and escape in the form of music.[3]

Listening to British Nature. Michael Guida, Oxford University Press. © Oxford University Press 2022.
DOI: 10.1093/oso/9780190085537.003.0006

For well over a decade BBC broadcasting had been an important part of disseminating and constructing facets of national identity in Britain, through both the act of tuning in with a nation of others and in the shared cultural effects of programming.[4] Siân Nicholas has shown that during WWII the practice of listening to the radio became central to creating a sense of unity and securing morale across the country.[5] It is in these contexts that birdsong took its place on air. The publications and broadcasts of Koch are the primary sources consulted here, alongside those of Nicholson and Huxley.[6] Documents from the BBC Written Archive, the *Radio Times* and contemporaneous writing about BBC broadcasting are also drawn upon.

Recording and modernizing birdsong

Until the mid-1930s, the everyday experience of hearing a bird sing could only be had in gardens and parks and perhaps from the kitchen or sitting-room windowsill in the spring and summertime. Ramblers heard a wide range of species in song, but many Britons did not have regular access to places outside the urban and suburban environments in which they lived and worked. Recording sound so that it could be reproduced on gramophone disc was a technique still confined to the studio. The BBC rarely recorded its broadcasts. An electro-magnetic device called the Blattnerphone that recorded onto steel wire and tape was installed at Savoy Hill in 1931, but it was difficult and expensive to use.[7] When in 1936 Desmond Hawkins broadcast his anthology of English poets who had written about English birds called *A Nest of Singing Birds* he had to call upon Imito, the Australian animal sound mimic, to bring the programme to life.[8]

There was in fact from 1911 a recording of a nightingale in the *His Master's Voice* catalogue that had been made by Carl Reich in Berlin. But this was a captive bird 'taken from its nest shortly after hatching, and reared by hand'.[9] Yet the recording was quite an achievement because nightingales were difficult birds to feed and they could quickly languish in captivity. To get a sound of sufficient volume and clarity to make an impression on the wax cylinder of his phonographic recorder Reich would have had to bring the horn of the device close to the cage without frightening the bird.[10] By 1914 the HMV catalogue listed seven of Reich's 'actual bird records': of a captive blackbird, a sprosser (a thrush nightingale), and a thrush, as well as a nightingale.[11] In 1927 recordings had been made and published commercially of the nightingale with Harrison's cello in her Surrey garden but they were not widely taken up by the public.[12]

Ludwig Koch's recordings

The first collection of wild British bird sounds to be recorded was made by a German Jewish émigré, exiled in Britain in 1936. Ludwig Koch arrived in London in the February mist and drizzle and immediately began to plan a recording project.[13] On a tip-off, Mary Adams, one of the BBC's first science programme producers, wrote to Koch asking to hear his collection of German birdsong recordings that he had been working on for years.[14] She was working with the anthropologist Tom Harrisson on a series of programmes about birdwatching, but Koch could not be persuaded.[15] He wrote in his memoir: 'I had to decline the offer since, with the approach of the nesting season, I wanted to start making new recordings which were to be the beginning of a new collection of British birds, and the editing of my continental records would have interfered with this.'[16] Walking one day in a college garden at Cambridge University, Koch had his first chance to listen to the spring song period. What he heard suggested to him that his work in Britain would not simply replicate his recordings of German birds: 'I had the impression that both the blackbird and the song-thrush sang more beautifully than I had heard them do in Germany.'[17] Were his exiled ears playing tricks on him?

Koch was a devoted and skilled nature sound recordist, who had worked as a director at Electrical and Musical Industries (EMI) in Germany where he was responsible for developing gramophone recordings for 'cultural' and 'educational' purposes.[18] It was here that Koch formulated the idea of a 'sound-book', comprising text, images, and gramophone recordings. He published eleven of these sets covering the sounds of animals recorded at Berlin zoo, birds in the woods around Brandenburg, and city soundscapes of Cologne and Leipzig, some recorded acoustically and others using the new Neumann electrical gear.[19] He published, with ornithologist Oscar Heinroth, his first sound-book dedicated to birdsong in 1935, *Gefiederte Meistersänger* (*Feathered Mastersingers*).[20] Using new developments in recording techniques, Koch was able to advance the limited conventions of the day that used paper and ink to make sense of birdsong for scientific or aesthetic reasons. He wanted to break away from the 'musical notations and curves which mean nothing either to a scientist or to a bird-lover'.[21] He felt, too, that the translation of bird sounds into words 'such as *tu, tu, tu* or *tse tse tse* will never bring to the ears of the average listener the sweetness of the song of the wood-lark or the characteristic note of the marsh-tit'.[22] Up to this point there had been no other ways to communicate the complexities and joys of birdsong, but for Koch too much was lost.

Koch's mission was to capture the songs of birds in their own habitats. It was worth the effort because he was convinced that caged birds did not sing in the same way as those in the wild.[23] His sound recording expertise had no doubt stemmed from his training as a pianist and then Lieder and opera singer.[24] He now sought out the big moments of bravura performance in birds. But to witness these and commit them to the unwieldy recording equipment available in 1936 was an enormous challenge of determination and technique (Figure 5.1). He chose the comfortable counties of Surrey and Kent to find his 'British' birds, taking with him a seven-tonne Parlophone recording truck and a small team of engineers.[25] The truck meant that he could not stray far from the road. Though Koch was usually recording very early in the morning, as dawn broke the sound of aircraft, trains, cars, and motorcycles often interfered with recording.[26] Apart

Figure 5.1 Ludwig Koch setting up a microphone with a colleague in a tree at night. (From Nicholson and Koch, *More Songs of Wild Birds*, 1937)

from the problems of the encroachment of modern noise was the challenge of catching a rare, fleeting moment when a bird was perched in song and in range of the microphones, before it took flight. In his truck, Koch would find himself listening to a bird having never seen it, yet the acousmatic thrill of a chance encounter could, with much patience, be inscribed onto wax. Even then, it is clear from Koch's explanation of the editing process that only tiny passages of recorded birdsong, 'often only a few millimetres', would go to make a record that the public would hear.[27]

Birdsong fixed for study

Using these painstaking processes, and knowledge of local hotspots of birdsong, Koch was able to fix on record, track-by-track, individual examples of the familiar fluttering life that so many Britons knew and loved—the blackbird, song-thrush, green woodpecker, nightingale, cuckoo, chaffinch, willow-warbler, whitethroat, great tit, robin, wren, hedge-sparrow, turtle dove, and wood pigeon. However well-known such birds were to the British public, listening experiences were only ever incidental and transitory for most. Rarely was a bird heard in full song, with all its variations, the listener given the opportunity to concentrate on those moments of individual performance. And who could identify a bird from its song? Koch believed that 'even farmers and woodsmen, who in spring often hear a song-thrush or a blackbird almost every minute, are often unable to name the singer'.[28] Birdsong was everywhere, yet still extraordinary and, in a way, unheard.

With Max Nicholson, who provided an expert ornithological text, and an introduction from the high-profile science popularizer and secretary of the Zoological Society of Great Britain, Julian Huxley, Koch's recordings were published in 1936 as a 'sound-book' called *Songs of Wild Birds*. This format, new to Britain, included a handbook with photographs, pull-out charts and two ten-inch gramophone records, all packaged in a colourful presentation box and selling for the costly sum of 15 shillings (Figure 5.2). This was a unique collection of British birdsong. The sound of the countryside had been distilled in birdsong and brought into the home. No other sound could best represent the entirety of British nature. To own the sound-book allowed the private listening of a public sound to take place in the comfort of one's armchair. On the cover of the presentation box of *Songs of Wild Birds* was a clean-cut man in dapper suit and tie studying the book while listening to one of the records. If the new sound technology appealed to men, the chance to listen to birds seemed to be targeted at them too. We should not worry that the man depicted may have been listening alone, for his pleasure was part of a burgeoning collective interest in bird life

Figure 5.2 Birdsong in a box. My copy of the sound-book *Songs of Wild Birds* (1936) showing the box cover and contents, which are a handbook and two 10-inch gramophone records.

in Britain in the 1930s that Nicholson was very much part of encouraging.[29] Nicholson was very active in widening the public's participation in scientific birdwatching. This sound-book collection, then, can be seen as part of a scientific and political interest in encouraging the British public to take an interest and learn about Britain's natural resources. Birds were emblematic of all of nature and their charm was a spur to pleasure, a prerequisite of a passion to learn. Huxley and Nicholson promoted this endeavour, but not only for its role in advancing knowledge of British nature. Huxley believed that birdsong was an 'expression' of the nation. 'The yellow-hammer's song seems the best possible expression of hot country roads in July', he wrote.[30] Knowledge of the yellow-hammer's song gave a sense of informed citizenship of the kind that understood the riches of the land. Knowledge acquired in the field or at home was a way of being a modern informed citizen who knew and valued their country.[31]

Koch, Huxley, and Nicholson saw this collection as a way for people get to know the character and complexities of individual bird sounds 'without distraction'.[32] A concentrated technique of listening would take the student beyond initial enchantments towards deeper appreciation and knowledge. The recording process allowed minute details to be noticed and heard in ways that were not possible in the wild. Listening was no longer bounded by an arbitrary moment; now it could be controlled and repeated. Moreover, the seeming realism of Koch's recordings of birdsong provided data with the virtue of disinterested objectivity, a quality prized by scientists.[33] Yet Koch had gone to great effort to detach the sound specimens of each bird from their natural environment, striving to isolate individuals as far as he could from the chatter of other birds. In this sense, Koch had produced a false impression of birdsong by removing the messiness of nature's sonic environment and the interactions of other birds. He had made efforts to avoid recording in countryside 'bird sanctuaries' as they would not have 'served our purpose of helping people distinguish particular songs'.[34] The collection was not intended to provide comforting rural atmospherics or bucolic montages for the urbanite. A second sound-book, *More Songs of Wild Birds* (1937), did offer a 'medley of bird's voices', however these recordings, which included a little owl, rook, jay and 'dawn choir', were there to afford 'a pleasurable exercise for the bird-lover who will have the opportunity of distinguishing one song from another'.[35]

The intention of these collections of birdsong was to educate the listener. This can be seen clearly in the way that Max Nicholson provided detailed listening notes as a guide to understand characteristics within and between species:

> (2 min 30 sec) The *cuckoo* is now heard calling. During the first fifteen seconds he utters a dozen cries without a break, then he flies away
> (2 min 45 sec) and is heard faintly for a few seconds in the background. Now at
> (2 min 55 sec) he is back again, and two other voices—a *chiffchaff's* and a *woodpigeon's*—can be heard faintly in the background.
> (3 min 00 sec) Now he is calling more deliberately, and in the first ten seconds of the third minute he only utters six notes. At this stage the *chiffchaff* becomes rather more distinct, singing a double note as monotonous in its own way as the cuckoo's, and just before the record ends a faint
> (3 min 11 sec) *blackbird* song is heard.[36]

This approach to listening, Nicholson argued, guarded against the risk of sentimentalizing the beauties of birdsong with 'false emotions or beliefs which might hinder a true appreciation'.[37] Distractions ruined the spell of bird music and full appreciation could be best had by listening in silence, seated comfortably with the lights dimmed.[38] Though Nicholson may have been asking too much of most people who bought these sound-book collections, we know that recorded birdsong could enable close listening and from this auditory state, both pleasure and knowledge were likely to flow.

For Huxley at least, the fidelity of Koch's recordings was evidenced by their ability to transport the listener to imaginative worlds. Huxley with his expert ear felt that to hear these records was 'to obtain a true picture of the birds' voices'.[39] It was this sense of the real, coming from the mediated, that Huxley said could evoke the singers and their natural environment:

As the nightingale's voice escaped from its ebonite prison under the touch of the needle and the scientific magic of the sound-box, I felt myself transported to dusk in an April copse-wood. The clear notes of the cuckoo with their blend of clear spring feeling and irritating monotony, the chaffinch's simple and cheerful strain, were equally evocatory; and with the laugh of the green woodpecker, the yellowing July fields and darkening green of July woods were in the room.[40]

Huxley could feel simultaneously a realness of the birds' presence in the room, but at the same time he found himself out in the woods. In both of these imaginative states, the medium had disappeared, a function of the faithfulness of the recordings as well as complicity in the listening subject to acoustic transparency.[41] I suspect that most who heard Koch's recordings were not quite as convinced. Certainly, within the home or study, the acoustics and physicality of being outside within nature was missing. The sensual impressions created by all senses working together were diminished indoors. And yet, if birdsong had been somewhat de-natured, tamed and rendered as a domestic commodity in its recording, it had also been privatized and brought close in a new kind of intimacy.[42]

The sound-books of birdsong sold well enough to surprise the publisher.[43] They continued to attract interest, two further impressions of the first being made during WWII, and many more over the next decade. Ornithologists loved these collections, but the sound-books were well-reviewed by the general press too and would have appealed to all kinds of bird lovers.[44] In 1936, *The Listener*, the high-brow magazine of the BBC, said that Koch's first set of discs offered 'a new vista of delight and knowledge to everyman'. Moreover, the reviewer felt there was something special about the sounds Koch had put on disc that distinguished them from the common currency of popular music: 'They are worth a dozen of the music everyone knows. They are worth twelve hundred cage-birds'.[45] These recordings were quite exceptional in Britain in the 1930s and formed the basis of Koch's BBC radio broadcasts throughout WWII. Before those broadcasts are considered, it is important to understand the kinds of listening attention that came with the anticipation of war and the new sounds of war itself on the home front. The way that birdsong would be heard, including Koch's recordings, would change in ways that meant these sounds became much more than sources of knowledge about the natural world.

Home front listening tensions

The anticipation of another war brought a vivid fear of aerial bombing. Public discussion of the threat began as early as 1932. In November of that year, Stanley Baldwin, serving in the coalition government, revealed his sense of the nation's vulnerability when he announced that there was no greater fear than 'fear of the air'. His prognosis was chilling: 'I think it is well also for the man in the street to realize that there is no power on earth that can protect him from being bombed'.[46] After only one raid, Bertrand Russell predicted, London 'will be one vast raving bedlam, the hospitals will be stormed'.[47] The threat from the air was sensationalized in films, novels, and political tracts.[48] The 1936 film *Things to Come* based on an H. G. Wells novel opens with mass chaos and panic in central London as an air attack begins on Christmas Eve. Anti-aircraft batteries fire into the night sky before bombs fall and destroy much of the city.[49] The soundtrack is the only education of the public imagination in the possible sounds of aerial bombardment should another war come. Before this was *All Quiet on the Western Front* (1930), the first major 'talkie' about WWI in which audiences could hear dialogue, as well the sounds of bullets, shells, and the screams of the wounded.[50] It mattered little that these sounds were simulated, whether they were over-done or under-done—they contributed to a fear of attack from the air.

The outbreak of war brought with it a preoccupation with the sky (Figure 5.3). Paul Nash, an official war artist for a second time, wrote these dramatic but telling words: 'But when the War came, suddenly the sky was upon us all like a huge hawk hovering, threatening. Everyone was searching the sky expecting some terror to fall.'[51] There were professional eyes and ears searching the sky when the Observer Corps were tasked with the job of identifying allied and enemy aircraft and collecting information in order to construct maps of aircraft movements for RAF Fighter Command.[52] But as Nash had indicated, public eyes were looking upwards at the same time. Several million citizens were armed with R. A. Saville-Sneath's book *Aircraft Recognition*, issued in February 1941 in the familiar orange Penguin paperback format.[53] Saville-Sneath gives a short chapter to acoustic identification, though he said that lack of experience of hearing enemy aircraft meant that recognition through sound alone was difficult.[54] However, another account of listening by the Observer Corps was confident that by knowing the sound of friendly aircraft, discrepancies could be detected as 'professional discrimination' was acquired. The Observer Corps found a newly revitalized sense of sonic mindedness:

> For many Observers the countless hours spent in listening have awakened a long dormant sense; the sense that registers, catalogues and above all, appreciates the infinitesimal sounds of which so many people are unaware. There is no such state as 'silence'. The Observer will never again be entirely lonely.[55]

This confident assertion of the power of careful listening by training the ear calls to mind the words of Koch and Huxley, who recommended the same process to reveal the richness of the acoustic world of nature. However, it was now part of

Figure 5.3 'Bren Gunners on the Alert against Air Attack' on the cover of the 23 May 1940 issue of *The Listener*, published by the BBC.

the war effort to pay attention to the sky to detect danger. In the propaganda films *London Can Take It!* (1940), the intended American audience was shown how in spite of the German Blitz on London, morale was higher than ever, while the narrator tells how 'listening crews are posted all the way from the coast to London to pick up the drone of the German planes' as teams positioned their listening apparatus.[56]

Listening out for danger was something many had a stake in. Ordinary civilians were listening in the cities, even during the daytime, as this passage from the journalist H. V. Morton makes clear:

Although the public appears to disregard daylight sirens, everyone is listening. Air war sharpens the ears, and a large part of the technique of self-preservation in raided cities is the ability to recognize instantly those sounds which are dangerous. It may be the first salvo of a nearby gun, or the swift whine of a descending bomb, which, although it may pitch half a mile away, always seems to be falling in a direct line to the crown of one's head. At such sounds, the streets empty.[57]

After raids, listening continued. Morton tells of a dog rescued that 'could be heard yelping in the ruins of a home'.[58] Rescue workers called for quiet to listen for people trapped in fallen buildings before demolition would go ahead.[59] Listening anxieties would reach their peak at night-time, when German bombing raids were most frequent. In bed, Virginia Woolf found herself 'listening to the zoom of a hornet which may at any moment sting you to death'.[60] Black-outs were a hated feature of the early home front war, Mass Observation reports indicated, their impact psychological as well as physical in villages as well as towns and cities.[61] The claustrophobia and practical inconvenience deeply affected the spirits. Home alone in the evenings, a typist from Liverpool was depressed and hedged-in by the blackness: 'I don't think I can go on like this all winter without going off my head', she reported in a Mass Observation survey.[62] 'One sits at home and drinks and smokes too much. One gets the wireless mania, too', reported a middle-aged male survey contributor from Runcorn. In the dark, there was the radio to listen to as well as the night sky overhead.

There was frustration too when attention to the sonic environment in wartime extended to restrictions of human speech, even silence being advocated as a condition of security and citizenship. People were asked by the authorities to 'keep mum' to avoid depressing morale by spreading rumours or passing on important intelligence. A raised consciousness of gossip, secrets, and spies lasted throughout the war, first through the Ministry of Information's 'Silent Column' campaign, then the 'Careless Talk Costs Lives' poster initiative. In the summer of 1944 the Daily Mail urged its readers to resist the temptation to speculate on the exciting progress in France, reminding them that careless talk was 'criminal folly' and must be curtailed.[63]

Radio talk was not curtailed, however—quite the opposite in fact. On the BBC, news airtime almost doubled during the war, while the main programming revolved around music and variety shows to lift the mood.[64] The wireless could counter the stresses of listening out for danger by providing practical information, news about the progress of the conflict, and the releasing effects of entertainment. There were problems in listening to the radio during air raids though. A BBC audience research report during the Blitz period of October 1940 found that in London and the Midlands listeners turned their set down low or switched them off completely, for fear that having it on 'may prevent them hearing the sirens or the noise of planes, bombs or guns'.[65] Yet, officially, the BBC's chief objective during this war was to bind the nation together as a community, and part of this effort was to give talk, comedy, variety, and popular and classical music a classless appeal.[66] Many talks in the first two years of the conflict, Siân Nicholas notes, focused on promoting Britain's cultural achievements and heritage, not least its 'English rural tradition'.[67] Music policy reflected an emphasis on broadcasting the well-loved English pastoral vernacular typified by Ralph Vaughan Williams, and while the popularity of 'There'll Always be an England' palled in time, 'The White Cliffs of Dover' was a consistent favourite.[68] As an 'instrument of war' the BBC Home Service became responsible for reflecting and asserting British national identity in order to sustain unity and morale.[69] All output would have to fulfil these objectives to some degree, including Ludwig Koch's birdsong programmes.

'Consoling voices of the air': Ludwig Koch's broadcasts

Koch did not begin broadcasting on the BBC until late 1941.[70] In the spring and summer of that year he was interned on the Isle of Man as an 'enemy alien', along with thousands of German and Austrian Jewish refugees. Yet he still managed to make a 'special study of the hooded crow and herring-gull' while there.[71] When released in August, Koch 'arrived back in London one midnight when a heavy air raid was in progress'.[72] When Koch's radio broadcasts began, the nation had already experienced the intensity of the Battle of Britain and the Blitz campaign, and the differing sensory textures of these two phases of aircraft dog-fights and city bombing, respectively.[73] Nobody knew what was coming next, so tensions did not necessarily dissipate, and after a period of relative calm, the summer of 1944 saw the jet-propelled V1 flying bombs, the so-called doodlebugs, deployed against London.[74] During the period in which Koch was broadcasting, well over 5,000 flying bombs were launched, more than 100 a day, and though many were shot down as expertise developed, 4,735 people were killed and 14,000 injured in and around London; 800,000 houses were damaged.[75] The artist Ceri Richards rendered in paint a British fighter plane intercepting a V1 flying bomb over the English Channel (Figure 5.4). In the turmoil of brush strokes is shown how thoroughly, at least in the mind, the sky had been newly occupied by raging flying machines.[76]

Working also in the air were Ludwig Koch's radio broadcasts of British birds in song, which I think served to reclaim the air that sirens, aeroplanes, and German bombs had set in frightening motion. Richard Overy has made clear the national aerial trauma of this time with the reminder that the violent death

Figure 5.4 *Falling Forms* by Ceri Richards (1944).

(© Estate of Ceri Richards. All rights reserved, DACS 2021. © Imperial War Museum, Art.IWM ART16504)

of over 43,000 people during the Blitz period alone 'was an unprecedented violation of British domestic life'.[77] Broadcasting's own kind of air power brought Koch's gentle interventions into homes during the late afternoon or early evening. His programmes reordered aerial awareness in adults and children and this can be glimpsed by the Observer's book series taking on the topic of 'airplanes' in 1942. The book took its place on the shelf next to British birds (1937), British wildflowers (1937), British butterflies (1938) and other natural history subjects, to inform the reader of important new aerial phenomena such as the silhouettes and engine configurations of the Bristol Beaufighter and Junkers 52/3M.

Koch was given regular slots on the radio from April 1941 and continued broadcasting throughout the war on *Country Magazine* and *Children's Hour* and with a series of five- to fifteen-minute solo shows. According to the *Radio Times*, Koch and his recordings appeared on air on thirty-two occasions during the war, most of which featured birdsong. *Country Magazine* was conceived as a wartime programme, which by 1946 had an audience of almost seven million listeners.[78]

The programme closed with one of Koch's 'sound-pictures' of the countryside.[79] *Children's Hour* had been running since radio began and went out during wartime just before tea-time. By appearing on *Children's Hour* Koch had the ears of millions, when many under-sixteens had been evacuated from cities and vulnerable coastal towns in the south and east of England. Popular with mothers and fathers listening with their children, the programme also acted as a daily point of contact between displaced family and friends.[80]

For his young listeners, Koch managed to convey the way he worked in the field—the early rising, the pursuit of a bird, and the exquisite pay-off of witnessing a good singer. This excerpt from a 1943 script gives a sense of how Koch would take his audience with him using vivid and involving descriptions to bring the outdoors to life:

> Let us creep closer, but carefully, not to do any harm to the birds' grass nests in a hollow in the ground. It is still pitch dark, before dawn; to hear the first lark you must get up with the lark. No noise, but many miles away in a hamlet I can hear faint barking and lowing. We are close to the skylark, it starts rising again. Here it is:—
>
> (Sky Lark)
>
> That was a very good performance and an extraordinarily good singer.[81]

Beyond these everyday moments, the status of Koch's birdsong broadcasts was indicated by their inclusion in Christmas Eve programming in 1941, when *Children's Hour* comprised a piece from Koch called 'Listen to Our Song-Birds in Winter', followed by a 'Visit to the Church of the Nativity in Bethlehem' and, finally, prayers.[82]

Koch went on to host five-, ten- and then fifteen-minute solo shows from 1943 to 1945. The 1943 series included *The Nuthatch Sings in February*, *The Mistle Thrush Sings in February* (with the bird singing with the backdrop of a thunderstorm), *Spring is in the Air* (with a nightingale by day, a cuckoo by night, the dawn chorus and garden warbler), and *The Song Thrush is Silent in August*. These programmes punctuated more conventional BBC programming that included talks, plays, light music, and the BBC orchestra. He combined a typically playful narrative with the educational in *The Nuthatch Sings in February*, broadcast on Sunday 7 February from 6:55–7pm: 'My particular nuthatch was living in a woodpecker's hole in a large chestnut tree. The entrance to the hole was too large for him so he narrowed it by filling it up with mud.' He then goes on to introduce the listener to six different nuthatch calls and songs, starting with the warning note ('it sounds, misleadingly, rather peaceful'), the angry call as Koch approached the tree, the mating note ('rather like boys whistling to each other'), and the trill which is sung 'on different levels, from a soft piano to a wide, carrying forte', and so on.[83] There was depth and detail in Koch's broadcasts as well as good-humoured entertainment.

If Koch's work was initially thought to be best suited to children's education and entertainment, as the war progressed Koch had the chance to address broader audiences, concentrating more on the solace and joy that listening to birdsong could bring, rather than the didactics. The evolution of Koch's output shows how the morale-boosting qualities of birdsong appear to have been increasingly recognized and given precedence over the pre-war focus on careful listening and the development of knowledge that his sound-books emphasized. His broadcasts move birdsong beyond a natural history audience to a much larger public in need of comfort and reassurance. When the tradition of the May-time live broadcasts of the Surrey nightingales was stopped in 1942, because the rumble of British bombers on their way to Germany was picked up by the microphones, Koch's recordings were all that remained of birdsong on the radio.[84]

Koch's reputation rose throughout the war.[85] Tom Harrisson, radio critic for the *Observer*, recognized Koch's work as special, linking it to John Grierson's much-respected documentary film work of the 1930s.[86] He complained that the BBC was using an impoverished repertoire of recorded sounds to enliven programmes: 'There is, for instance, a snatch of B.B.C. Seagull, which I have heard represent "the sea" over and over again—once four times in a day.' He urged senior executives at the BBC to pay attention to Koch's 'aural documentary' collection in this way:

> listen to his sweet, babbling brook; his dying bumble bee; the terrifying 'Symphony of Steel', and clever Victory V; the strange chaos of his regiment drilling, and beautiful dawn chorus of birds. Listening to these unused records one hears a whole new area of radio reality.[87]

Harrisson credited Koch for using his recording expertise as an ethnographic method of documenting Britain's military, industry, cities, and countryside at war. 'Radio reality' for Harrisson referred to a broadcasting that fired the imagination by bringing authentic national sounds and voices to the listener in sonic tableaux, not just as discrete sound effects.[88] Koch's work, then, could represent the nation through its honest depictions of everyday identities.

Next will be an analysis of the functions and reception of these programmes. What social and cultural work did they do and for whom? There are two parts to this. First, I argue that Koch's birdsong broadcasts provided consolation via the sound of ordinary birds that everyone could appreciate. Koch's recordings of birds were no longer limited to private consumption; these sounds of national heritage had become public property again through their broadcast. Second,

I develop an argument about how Koch's birdsong was seen to speak of a soft and gentle patriotism and a natural fortitude, rather than a militaristic one, rooted in notions of citizenship built on knowing the land and its creatures.

Solace for all in 'classless' songs

Two weeks before the evacuation of Allied forces from the beaches of Dunkirk in 1940, a letter from Koch appeared in *The Times*. Koch encouraged readers to find comfort in the beauty and persistence of birdsong:

> War or no war, bird life is going on and even the armed power of the three dictators cannot prevent it. I would like to advise everybody in a position to do so, to relax his nerves, in listening to the songs, now so beautiful, of the British birds.[89]

From this we learn that before he began broadcasting, Koch was already convinced that the sound of birds would help people cope with the crisis. It is interesting that Koch mentions that hearing birdsong might be good for the nerves. A Mass Observation report had confirmed that the nerves were affected for the worse by the sounds of attack: 'It is the siren or the whistle or the explosion or the drone—these are the things that terrify. Fear seems to come to us most of all through our sense of hearing.'[90]

During the 1930s the BBC had placed a distinct emphasis on the consoling power of radio, seeing it as an instrument of solace, not simply a broker of ideas, culture and entertainment.[91] The combination of this vision of broadcasting's gentleness and Koch's sounds of British nature reinforce one another during the war. Koch went so far as to feel that recording and broadcasting the timeless sounds of the farmyard would be a direct challenge to the sound of German bombs:

> I was allowed [by the BBC] to make all kinds of recordings. I visited a number of factories to explore unusual noises, but amid the din of machinery I longed for the sounds of nature, and persuaded my superiors that this was the right moment to show the enemy, by recording all kinds of farm animals, that even bombing could not entirely shatter the natural peace of this island.[92]

Koch's eccentric plan for a programme does not seem to have been aired. His intention though was to bring the comfort of the pastoral into the front room. The British Library holds a carefully mixed four-minute montage, which is likely to have been the basis for a broadcast. One hears the voices in close up of cockerels,

hens, sparrows, ducks, geese, a donkey, a horse, piglets and pigs, cows, and a concluding sequence of sheep and lambs bleating in bucolic chorus.[93]

The 'natural peace' that Koch spoke of was to be found in the countryside,
where so many evacuees had been sent, a place that before the war had been
enjoyed by increasing numbers of ramblers and youth hostellers at the weekends.
Now mobility was severely restricted by petrol rationing and rail disruptions,
reflected in the government policy idea of 'Holidays at Home'.[94] A letter in the
Radio Times from Dora Read in west London in 1943 underlined how Koch's
broadcasts connected listeners with the peace of the countryside that was now
missed: 'Many thanks for letting us hear the wonderful birdsong, full of hope and
peace to come. Millions of us, used to rambling before the war, are now in factories doing war work. Let us hear more of Ludwig Koch's birds!'[95]

Reflecting in his 1955 memoir on his public audience, Koch believed that
his birdsong programmes had piqued interest across all lines of age, class, and
gender: 'among my listeners there are obviously a great number of adepts, men
and women of all ages, and of all classes of society'.[96] Koch's sense of self-impor-
tance shines through here but Julian Huxley, too, saw birdsong as an egalitarian
joy that anyone could appreciate. 'I suppose that birds give more pleasure and
interest to humanity [. . .] than all the other groups of the animal kingdom taken
together', Huxley wrote in his introduction to Koch's first sound-book. Huxley
continued, saying that birds through 'their beauty, vivacity, by their songs and
freedom of flight, by their migrations and their domestic arrangements, they
make an obvious appeal to the layman, however uninstructed'.[97] Huxley argued
for the benefits of attentive listening for the acquisition of knowledge, but he also
acknowledged that this was not a precondition for enjoyment.

Huxley went further, arguing that the sense of hearing accessed and seduced
the emotions in ways that intellectual engagement could not: 'The associations
called up by sound seem to share with those aroused by smell the properties of
fullness, immediacy, and emotional completeness to an extent not aroused by
those dependent on sight or intellectual comprehension'.[98] This is a significant
thought—if sensory perception rather than intellectual reflection was what was
needed to get the most out of listening to birdsong, this meant its pleasures were
open to all Britons. Such a notion would have fitted with the tendency during
the war for the BBC to move away from the cultural elitism of its programming,
towards an uneasy kind of 'elevated classlessness'.[99]

It is certainly true that what listeners heard in Koch's selection of birds were
ordinary voices, some of which would be familiar even if they could not be identified precisely. One can argue that the songs of the robin, blackbird, cuckoo, and
nightingale were part of a sonic national character rooted in the everyday ordinariness that became an increasing part of the tone of broadcasting during the
war. J. B. Priestley's enormously popular *Postscripts*, broadcast at the height of

invasion fear, brought his mellow Yorkshire accent and down-to-earth manner to millions. Propaganda was most successful when it adopted a tone that chimed with typical British citizens.[100] In addition, from 1940 there was an effort to get regional dialects and class accents into talks and features and light entertainment, while announcers with northern English and Scottish accents were given the microphone.[101] Koch's voice was anything but ordinary. We don't know how it was interpreted by radio listeners, but it is rarely mentioned. We can assume though that it came across as interestingly foreign rather than obviously German.[102] His eccentric, high-pitched delivery was unmistakable, and with his heavy German accent he would almost sing the English language.[103] In his broadcasts ordinary avian voices were framed by an extraordinary human one.

In February 1940, John Reith, no longer at the BBC, but knighted and Minister of Information, advanced the idea that there was a need for new national songs that everyone could share and get behind. He suggested approaching the nation's leading composers to write a series of what he thought could be 'lay hymns' of 'the Jerusalem brand'. Reith wanted patriotic themes 'but not necessarily war-like' ones.[104] The suggestion does not seem to have been successful, but, in light of this, one might see the birds singing from the wireless as the composers of small pastoral lay hymns that everyone could enjoy. These birds could even be said to have been providing their own wordless national anthems in miniature, ones that reminded radio listeners of Blake's green and pleasant land. Vera Lynn's romantic wartime tune, 'A Nightingale Sang in Berkeley Square', made use of the idea of an ancient motif from nature within a new national song. Patriotism was undoubtedly read into British birdsong in this period.

Patriotic songs

Koch's broadcasts can be seen as part of explicit patriotic celebrations of rural heritage. *Country Magazine*, a programme about the vitality of the countryside and its farming communities that Koch featured on, was launched in 1942. It was accompanied by others, including *The Countryman in Wartime* and *Your Garden in Wartime*. Another programme, *The Land We Defend*, pictured Britain as one vast and pretty village populated by lovers of nature and countryside.[105] The English countryside, which stood for peace, tranquillity, stability, harmony, and timelessness, was emblematic of what Sonya Rose has referred to as the 'authentic nation' during WWII.[106]

However, Koch's programmes did more than refer to ideals of a romanticized pastoral southern 'England' of the past, where he had recorded his birds but where few lived.[107] Birds and their song had a place in many British lives, including suburban and urban ones. Birds were not simply a symbol of rolling green landscape and country lanes, the one that recruitment posters had employed in both wars, so much as a vibrant and present reality in people's lives.[108] Birdsong was not simply a cue for nostalgic longing for a lost past; it could point to the newly built suburbs with their little gardens and access to other green amenities that enabled, in Priestley's words, 'the salesman or the clerk, out of hours, to be almost a country gentleman'.[109] There is a residue of the 'deep England' myth in Priestley's words, but the point is that to hear birdsong was more to do with everyday reality than everyday fantasy. Birdsong was more than a symbol of British nature; for many it *was* British nature because it called attention to itself more than anything else. The link between birdsong and national identity was put to use in the propaganda film *Listen to Britain*, which defined national character through the chatter of birds, the rustle of summer corn and peaceful rural scenes, together with the powerful modern sounds of British aircraft, factories and coal mining.[110] All these sounds were indicative of a modern Britain that could win the war.

Koch was by no means the only enthusiast for birds and their song who was active and vocal during WWII. Books about birds were published quite intentionally in the midst of war (and the *Observer's Book of British Birds* was reprinted at least eight times). One small Pelican paperback placed great emphasis on the belief that paying attention to birds could improve the lives of ordinary people at war.[111] Ornithologist James Fisher's book was called simply *Watching Birds*. Writing just after the Battle of Britain, in November 1940, Fisher placed birds at the centre of the conflict:

> Some people might consider an apology necessary for the appearance of a book about birds at a time when Britain is fighting for its own and many other lives. I make no such apology. Birds are part of the heritage we are fighting for. After this war ordinary people are going to have a better time than they have had; they are going to get about more[...] many will get the opportunity hitherto sought in vain, of watching wild creatures and making discoveries about them. It is for these men and women, and not the privileged few to whom ornithology has been an indulgence, that I have written this little book.[112]

Birds were part of the nation's heritage and identity, threatened by invasion from the sky or the coast, and a precious natural resource for the future, Fisher argued. While Fisher's book is a serious work of ornithology covering anatomy, migration, habitats, territory, and courtship, with technical illustrations and charts, it went on to sell over three million copies and is credited for enthusing a whole generation of the public into an appreciation of birds.[113] Perhaps to possess such a book, without getting too involved in the detail, allowed the owner to be invested in something of the nation's bird heritage and its consolations.

Patriotism, but also scientific interests, were at play when British birds were declared the best singers. Koch had demonstrated on air in 1944 during *The Song Thrush is Silent in August* 'the great superiority of the British over the German song-thrush whom I also know well'. He played first his German recording, then his British one, asking the listener to make up their own mind.[114] His refugee status in the safety of Britain may well have influenced how he heard British birds, but he was not the only one who held such views. Seasoned ornithologists like Max Nicholson had made similar claims in the 1930s, asserting that in no other country was birdsong as powerful, varied, and pleasing as in England. The fact that so many resident species were 'good songsters', common to gardens and familiar to ordinary people, made 'England a paradise for bird-song'.[115]

One further kind of patriotic spirit, reflected in the short-lived 'Keep Calm and Carry On' poster campaign, was observed in the behaviour of birds. The Ministry of Information's slogan drew its inspiration from public and private discourse during WWI.[116] The aim of the injunction was to encourage wartime resilience, particularly qualities of fortitude on the home front. So it was in birds, too, that a kind of Blitz-spirit was recognized, embodied in their vocal performances amid the noise and chaos of conflict. Such apparent endurance seemed to provide inspiration and hope that birdlife, and therefore human life, would prevail. It was part of the myth-making that was prevalent and needed during the war. Ludwig Koch boasted that 'a Spitfire's drone would not scare a nightingale',[117] although the naturalist Richard Fitter reported that 'one result of the "fly blitz" of 1944 was to drive many of the woodpigeons from the London parks'.[118] Not all birds seemed to display Blitz-spirit, but Fitter's 1945 book *London's Natural History* is nevertheless a notable tribute to London's wildlife as robust and irrepressible, rising out of the bombsites.

Birdsong civilized and civilizing

Mixed with the patriotic stories that Britons told each other about the superiority and durability of their country and its creatures were more fundamental thoughts: that humanity needed the close association of birdlife to be fully civilized, to flourish and to progress.[119] In birdsong, Koch, Huxley, Nicholson and others heard the civilized markers of long-proven social harmony, moral conduct, and artistic musical culture. Avian society appeared to possess all these things, although different parts of elite discourse emphasized particular qualities. The ideas were sometimes only part-formed and they were unorthodox in that concepts of Western urban civilization have usually involved separation from and dominion over the natural environment by cultural elites.[120] However, it is possible to argue that birds and their song suggested to the key characters

considered here some of the ideals of refined human social development, when
the state of civilization was under question during WWII. I will go on to show
how birds themselves could be looked upon as civilized beings and how in turn
they could even be capable of civilizing the human barbarism of the Nazi enemy.

In 1930, Julian Huxley had published *Bird-Watching and Bird Behaviour*,
based on a six-part broadcast series earlier in the year. In its final chapter, he
explains the unique evolutionary trajectory of birds that has made them so different to other living things, but then sums up with a plea that birdlife deserves a
privileged place next to the human:

> For—and this is my last word—in considering the birds' place in Nature we
> must remember that they have a place in civilization as well as in wild na-
> ture, and that even if we be mechanizing so many aspects of life, or rather, just
> because we *are* mechanising them, there is all the more reason to reserve to
> birds—shy birds as well as tame, rare birds as well as common—a place in our
> civilized scheme of things, and that to see that that place is kept for them, and so
> for our delectation and that of our posterity.[121]

Having acknowledged that bird life is rather special and distinct in the way it
has evolved, Huxley wonders how Britons will manage modern life without the
company of birds great and small around them. He had noticed how town and
city life was enhanced by the presence of birds in America and Germany, where
he had seen bird-boxes, bird tables, and bird baths successfully encouraging bird
populations to be part of the social life of urban-dwellers.[122] Huxley's reference to
posterity indicated a role for birdlife in human lives, beyond everyday pleasures,
in the sustenance of new generations facing the continued pressures of industrial
modernity. He felt that humans could best survive and evolve in the company
of birds, as if both parties would benefit by some kind of ecological and evolutionary synergy. Huxley had held a professorship in zoology at King's College,
London, but given it up in order to collaborate with H. G. Wells and son to write
what became the most widely distributed account of the new Darwinism, *The
Science of Life* (1929).[123] While Huxley was the foremost public communicator
of evolutionary science at this time, this did not prevent him from taking these
ideas into more speculative terrain when thinking about the relationship between societies of living things.

In more prosaic ways, Huxley found birds to be excellent companions to the
rambler or hiker: 'To go out on a country walk and see and hear different wild
birds is thus to the birdwatcher rather like running across a number of familiar
neighbours, local characters, or old acquaintances.'[124] Some knowledge of birdsong, he says, 'makes each kind of bird a single and perennial friend'. In other
words, by listening, the chorus of birdsong is differentiated into individuals with

their own identities. One no longer travelled alone. In this sense, birds were a part of the fabric of civil society that could be relied upon to enhance a wider sense of community with their rich and diverse personalities.

As a second war became more likely, Max Nicholson also looked to the singing of birds as a reassurance that civilized life would continue. In 1936 he wrote these lines in the book that accompanied Koch's first set of recordings of British birds:

> In a world of growing complexity and difficulty we turn to bird-song as something which is not only delightful in itself, but which has hardly the remotest connection with human worries. We may be uncertain whether London and Paris and Berlin will be reduced to heaps of ruins by the misuse of scientific weapons in the interests of mutual destruction, but we can be sure that in any case nightingales will sing in Surrey every May, and golden orioles will still flute with civilised perfection in German and French spinneys, regardless of human barbarism or of human achievements.[125]

This is a humanist statement not a nationalistic one—birds of *all* nations, Nicholson says here, could be relied upon to carry forward civilization, however unreliable and quarrelsome human relations proved to be. He also intimates that the rarefied sensibilities of birds were in the end superior to those of men; birds flew above human chaos, they sang out with 'perfection' in spite of war, their straightforward moral and aesthetic senses pure, their communication honest and heartfelt. Huxley saw birds as participants and partners in the scheme of modern civilization that would be impoverished and isolated without them. Nicholson felt that birds were in some sense more moral beings than humans, and were quite capable of carrying forward their own kind of cultured and ordered society alone.

As a musician and singer, Ludwig Koch had his own opinions about the cultural capacities of birds. For him, they were the most 'artful of all living things'.[126] Birds produced music of superior value than the popular tunes on the radio, a critic in *The Listener* magazine had argued on hearing Koch's recordings.[127] Koch saw himself as a collector of the best musical performances of birds, but he also intended to document many other characteristic elements of Britain's sonic heritage—natural history, folk-songs, dialects, the voices of famous men and women, the distinctive sounds of different industries and cities. Koch's ethnographic approach to documenting the world in sound brought together birdlife with human life. Ultimately he envisioned all these recordings being housed in a 'Sound Institute' with local listening branches around the country.[128] Though this idea was not realized, Julian Huxley, as the first director general of UNESCO, resolved in 1948 to have Koch's collection of more than 500 wildlife recordings preserved.[129] This material was both a national treasure and a contribution to

UNESCO's objective of promoting international understanding, cooperation, and peace, because it demonstrated a part of the distinctive quality of British culture.[130] If birdsong was a constituent of British culture then it must also have been part of its story of civilization.

Civilization in crisis?

The kind of thinking we have examined here can be better understood if we cast an eye back to WWI and its aftermath. Historians have often viewed that conflict as a trigger for the contemporary reassessment of human progress. British servicemen and women, after the victory in 1918, were given a medal that stated they had fought in 'The Great War for Civilisation'. Yet the reality of that war, more than any other, seemed like the utter negation of civilized values—humans reduced to the level of animals in the mud, blown to pieces by modern machinery. Civilization in all its sophistication had undone itself.[131] In the shadow of war, the future of Western societies was debated intensely by intellectuals in Europe and America throughout the 1920s.[132] At home this was much more than a debate about how Britain and the allied powers had defeated German brutality. Britain had its own fears of social decline or collapse, and even if they were often elaborated in ways that defied historical reality, talk of 'civilization in crisis' became something of a cliché in the interwar years. Richard Overy has pointed out that the morbid moods stemmed not just from rhetoric but were rooted in serious scientific, medical, economic, and cultural descriptions of the present.[133] Other discourses existed, certainly, but 'pessimism was highly contagious', Overy has argued.[134] The Egyptologist Flinders Petrie, with his *Revolutions of Civilisation* (1911) and Arnold Toynbee's broadcasts in 1931 followed by his *Study of History*, the first part of which appeared in 1934, familiarized many with the idea that civilizations rise and fall, rather than constitute a narrative of sequential progress. Oswald Spengler's *The Decline of the West*, once it was translated and issued in the early 1920s in Britain, never sold particularly well, but the title became popular shorthand for cultural pessimism.[135]

The remedies to the apparent crisis in the 1920s and 1930s were many, and Max Nicholson had a role here with his work in economic planning, a project seen to be a primary solution to interwar forecasts of peril.[136] Together with his re-thinking of the practice of ornithology and his leadership to establish the British Trust for Ornithology, Nicholson was in 1930 the assistant editor and leader-writer for the new political weekly *Week-End Review*. He authored a special report in the *Review* called 'A National Plan for Great Britain', which was the stimulus for the formation of the liberal grouping called Political and Economic Planning (PEP).[137] Led by Sir Basil Blackett, a former government Treasury civil servant and director of the Bank of England, Labour politician Kenneth Lindsay, and Israel Sieff, vice-chairman of Marks & Spencer, with Julian Huxley responsible for directing research, PEP sought to use planning methods to rescue the existing order from an impending social and economic disaster.[138]

Nicholson's National Plan attacked the 'present chaotic economic and social order' and in a following report he wrote that 'the anarchy and squalor of Western civilisation has come to a head'. The purpose of planning was 'to reconcile personal freedom with an orderly community'.[139] In other words, human society could not be left to its own devices to function properly. Nicholson, unsurprisingly, does not bring birds into his PEP theory, though it is clear that the state of human civilization after WWI prompted Nicholson to see birdlife as an example of a society at peace with itself. For Nicholson, and Huxley as well, birdlife was a stable *model* of nature, with in-built evolutionary and ecological checks. Unregulated human economics created cycles that tended towards failure, while the natural laws governing birdlife were manifestly successful in creating order, stability, balance and a harmony signified by song. While Nicholson and Huxley thought that a planned economy would be necessary to avert a human disaster, they also had a sense that humans would be better off if their civilization included birds. By 1936, when another war was seen as a certainty by many, and Nicholson and Huxley were collaborating with Ludwig Koch on *Songs of Wild Birds*, they perceived a message of hope for a joint animal-human collective in the song-making of birds. There was little doubt that birds could carry on without humans in the world, but could birds help human civilization to continue?

Broadcasting civilization

The process of broadcasting was itself supposed to be civilizing—John Reith had seen it as 'part of the permanent and essential machinery of civilisation'.[140] BBC broadcasting's duty of service was to put before the public the very finest cultural exhibits. One particular broadcast story from the wartime period allows for an exploration of the potent place of birdsong in influential British minds, when many sought out solutions to avert the crisis. Harman Grisewood was assistant director of programme planning at the BBC in 1939 when he visited Berlin just before the outbreak of war and was sickened by the 'yellow-marked benches for the Jews in the Tiergarten'.[141] On his return he went to see the then director general of the BBC, Frederick Ogilvie, to report on his impressions. Grisewood recalls:

What he said was terrifying; I can still remember it word for word. He said: 'You know the Germans are very sentimental people.' I said: 'Yes it's often explained to one that this is so.' He then said: 'Well, what we're going to do is broadcast the nightingale to the Germans. The cellist Beatrice Harrison will go into the woods near Oxford and play her cello. The nightingale will sing and we'll broadcast that to the Germans.' I felt there was no point really in going on with the conversation.[142]

Grisewood felt that Ogilvie could not or would not digest the 'horrible proportions' of the situation, so instead had suggested that the broadcast to the Germans would be 'a token of our peace-loving intentions'.[143] Grisewood was uncertain how serious Ogilvie really was, but several broadcasting scholars have cited this report as evidence within the BBC of narrow-minded complacency and a tendency towards appeasement in 1939.[144] British officials found it very difficult to believe the accounts of Jewish persecution told by refugees fleeing Germany, Tim Crook has argued.[145] In Ogilvie's appeasement plan there was a hope that if human diplomacy had not worked then this unique voice of nature that had moved so many British hearts on the wireless each spring for over a decade might be worth a try. It was a desperate hope from a very powerful man and perhaps we must assume that Ogilvie did not really believe the song of the nightingale would soften hard Nazi hearts and persuade them to take a more peaceful view. But we can also assume that he was confident that such a broadcast, should it happen, would encapsulate all that was good about Britain, its values, its people. It would not show weakness, rather the shared 'sentimental' character of the German and British people. It would be a cultural exchange between the two countries that could not be encapsulated in words, an attempt to communicate an essence of the British national character in bird and human music. The performance of bird and cellist would be an emphatic reminder of British civilized values, even if it could not move the Germans who listened.

If the nightingale in song with Harrison's cello been broadcast to Germany it would have echoed something of the English mind and humour that J. B. Priestley had described as 'rather temperate and hazy', 'blurred and kindly'.[146] The imagery of English character that was crafted during WWII in the broadcasts of Priestley and Noel Coward, and by George Orwell in his essays, prompted Raphael Samuel to comment that in their self-image 'the English were a domestic people rather than a master race, home-lovers rather than conquerors. Their patriotism was quiet'.[147]

This wartime mythologizing of a peaceable and soft-spoken national character—though *England* was usually the reference point—was set against an equally simplistic depiction of what Brian Currid has called the 'sonic icon' of Hitler's shouting voice and the deafening crowds at party rallies (Figure 5.5).[148] Priestley said in his *Postscripts* broadcasts that the Nazi 'loves bluster and swagger uniforms and bodyguards and fast cars, plotting in back rooms, shouting and

Figure 5.5 The book of the BBC broadcast series 'The Voice of the Nazi', in which the softly spoken professor of psychology W. A. Sinclair attempted to expose the falsehoods behind Nazi rhetoric.
(Published by Collins, 1940)

bullying, taking it out of all the people who have made him feel inferior'.[149] A *Daily Mail* columnist argued that Britons who listened to Hitler's voice, due to be broadcast that day, would hear a 'merciless use of voice, and though some may not understand a word he says, still they will be able to identify that tone of menace, that barking attack on the senses'.[150] How much of or how often Hitler's voice was heard on the radio by the British public is unclear, but in preparation for the 1939 BBC drama *The Shadow of the Swastika*, recordings of the speeches of Hitler, Goebbels, and Göring were studied by the British cast to create 'exactly the tone of voice and emotional feeling' that would bring home the essential nature of the Nazi movement.[151]

Priestley's Sunday evening *Postscripts* talks during the summer of 1940 are as inseparable as Churchill's speeches from the national mood at that critical period.[152] His talks remind us how the British brand of civilization was conceived. Key was his way of talking about 'all of us ordinary people', most famously illustrated by the 'people's war' vision elaborated in his 5 June talk about the Dunkirk little ships.[153] Everyday ordinariness was embodied in the tones and virtues of Britons: 'simple, kindly, humorous, brave', 'imaginative and romantic', and possessing 'courage and resolution and cheerfulness'.[154] Complementary to these human qualities was the everyday birdsong that Priestley heard at the end of one of the loveliest springs he could remember: 'Just outside my study, there are a couple of blackbirds who think they're still in the Garden of Eden.'[155] Priestley's whimsical words also reflected the image of civility long believed to be demonstrated in good quality conversation.[156] The chatter of birds was heard as conversation between sophisticated, well-mannered creatures. When he encounters one dark night in a Hampstead pond a quacking duck and its 'faint squeaking' ducklings, he tells his listeners that he is able to understand the war as a battle between despair and hope:

> For Nazism is really the most violent expression of the despair in the modern world. It's the black abyss at the end of a wrong road. It's a negation of the good life. It is at heart death-worship. But there flows through all nature a tide of being, a creative energy that at every moment challenges and contradicts this death-worship of despairing crazy men.[157]

The mallard family had crystallized the gulf between the civilized and the savage, between the vibrancy of nature and oblivion. Priestley had in another broadcast contrasted British virtues with those of the Nazi enemy, who were dehumanized and robotic: 'thin-lipped, cold eyed', 'a kind of overgrown species of warrior-ant'.[158] This was the dark side of modernity, modernity out of control, a civilization gone wrong. Britain was proudly modern but had not abandoned its values and traditions. Priestley told his listeners that those blackbirds outside his window would be part of Britain's future. They had been there 'long before the Germans went mad, and will be there when that madness is only remembered as an old nightmare'.[159]

Conclusion

In 1936, for the first time, a collection of British birdsong was recorded onto shellac. In fact, it was song from the 'south country' of England, as Edward

Thomas had called it, where familiar friends like the blackbird, thrush, chiff-chaff, robin and wren resided. A nightingale was recorded as well, and this sound was unknown north of the Midlands. All could be savoured at home when Ludwig Koch released these iconic sounds of nature packaged in a sound-book. They were the first wild sounds to be recorded and distributed for public consumption. Now the fleeting complexities of a bird's voice in the field had been inscribed to give human perception and appreciation another chance. These sounds had been tamed, and could be owned and studied indoors. This was not the first time that birdsong had been brought into the front room, how-ever—for a century wild birds were caught and caged for the company of their song, and cotton towns of Lancashire and mining communities in Yorkshire and the Scottish lowlands had vibrant canary fancies.[160] If owning a gramophone record was less exciting than having a bird of one's own, it did mean that the intricacies of birdsong could be listened to, contemplated, and understood in new ways. Birdsong had been modernized in that this fixing of notes allowed and encouraged concentrated listening, over and over, in season or out, and thus, birdsong could be known better.

Koch's BBC broadcasts of birdsong throughout WWII made these small private sounds part of public culture, food for a nation at war. As such, Koch's birdsong programmes—gentle, commonplace evocations of precious natural heritage under threat—became a state tool for prosecuting the war. Yet they were free from the clichés of nostalgic ruralist propaganda that radio listeners reacted against, because these birds actively revitalized the air when the skies were heavy with danger. Koch's birds were alive in ways that the picturesque landscape, its sleepy cottages and lanes, were not. Birds sang out clear and defiant in the con-flict; they carried on as all citizens had to. As Koch gained in popularity, he was allowed to move beyond his initial *Children's Hour* audience to address the na-tion in prime weekend slots. Birdsong united listeners, when radio music and talk often divided opinion.

With human conduct in war under question, J. B. Priestley told listeners to his programmes that bird voices were reminiscent of the softly spoken civilized modes of British behaviour. For Koch, birdsong was a kind of music, produced by creatures with refined aesthetic sensibilities. In the modernizing world, and in one threatened by human barbarism, both Julian Huxley and Max Nicholson argued that bird life should be considered part of civilization; humans might gain inspiration from their beauty and conduct. On the eve of war, the potency of the nightingale's song reappears when the BBC's director general Frederick Ogilvie wonders if the bird singing again with Beatrice Harrison might persuade the Nazis to take a more peaceable view. The idea was a return to the wonderment of the live transmission, but most of all a demonstration of how closely British iden-tity and birdsong were linked.

Afterword

The sensing of the natural world in the first half of the twentieth century was the business of soldiers, school children, broadcasters, doctors, ramblers, and many others in British society. Almost everyone had a reason to seek out some kind of pleasure in their natural surroundings, not least the urban many. When people listened to nature, they were selecting and ordering the many overlapping sonic textures that made up modern life, and in doing so made sense of their own lives. Listening was not like looking—it required careful engagement with one's changing environment, ever subject to the flit of time. Listening accounted for the spherical space that extended all around. Listening to nature was a way of drawing close its presence, so valued in times of crisis and change.

And yet there was then, and there remains today, a nervousness that dwelling on nature's sounds and sensualities became too easily a nostalgic indulgence, that it came with a denial of the realities of modernity, that it could not inform what it meant to be modern. However, in 1924 the essence of English national character could still be located in the sounds of the smithy, scythe, and corncrake, according to Stanley Baldwin, who led the Conservative government of the United Kingdom and its empire three times between the wars. Baldwin could successfully claim these traditional sounds to be English and relevant because they were companions to and supportive of a multifaceted industrial prosperity. Almost a generation later, just after another world war, George Orwell expressed frustration in his essay *Some Thoughts on the Common Toad* that many considered 'any pleasure in the actual process of life encourages a sort of political quietism' and brought charges of sentimentalism.[1] For both Baldwin and Orwell, paying attention to the nation's natural world and identifying it as special was not incompatible with their very different visions of what Britain was and could be. Without the soft buzz of nature in train, progress could not be envisioned securely.

The nervousness of assessing the appreciation of the natural world in everyday culture is evident in today's scholarship of sound in society, which has favoured, rather, analyses of the roar of the city, the experience of living with new machines, and the emergence of new musical forms. This book has found that the sounds of nature were ever present in British society, at peace and at war, and that they were used to gauge the status of a world in flux. R. Murray Schafer's influential ecological sonic hierarchy that demonized and then excluded the human from the ideal 'soundscape' created a model of sonic modernity that would not accommodate both man-made and natural sounds in concert. The

Listening to British Nature. Michael Guida, Oxford University Press. © Oxford University Press 2022.
DOI: 10.1093/oso/9780190085537.003.0007

reality of auditory experience in this study, however, has been a messy mix of the newest and the oldest sounds from which Britons have had to make and remake their own meanings. From this mix all Britons to a greater or lesser extent found the sounds of nature in their lives. These experiences have not been confined to children (who might also learn language via animal sounds), or to poets (who first notated birdsong on the page). Nor has their thrill spoken only to composers of pastoral music or to naturalists. The invigorating sounds of the natural world were freely available to everyone and cost nothing. They pressed their way even into the most shaded corners of inner city existence.

This work has added to the understanding of how men protected themselves and survived in the Western Front trenches by knowing as well as imagining what the sounds of nature around them meant. It brings into the open the use of pastoral quietude as a therapeutic milieu to aid recovery from the psychological shock of war. Unravelled are the ways in which the idea of a silent nature was incorporated into early BBC broadcasting, to soften the threat of its new voice but also to suggest its transcendent potential. A new reading of the culture of interwar rambling goes beyond considerations of regional and national political identities to emphasize the importance of sonic and sensual alternatives to urban modes of existence. And during WWII the continued cultural potency of bird-song, even when recorded and broadcast, is connected to ideals of national character and civilization that included conversations with the natural world.

If modern life in the years of 1914 to 1945 often kept the natural world at a distance, it also created conditions that meant humans needed its proximity and paid new attention to its energies. The soundscapes of wartime and of new technologies did not obliterate the sounds that nature made, rather they established a contemporary auditory framework in which the natural world was heard afresh. Modern lives were drawn to these primal sounds because their ancient authority was recognized. They were listened out for—sought out—by urbanites who found the sensations of their towns and cities limited and dissatisfying. Birdsong was heard as a symbolic call from the wild, at the same time as it was part of the everyday for many Britons. To witness the song of a blackbird was to hear a new beginning. Its garrulous energy was understood as a signal of renewal and refreshment and humans were taken onwards by the flow of notes. The language of birds was heard as intricate, musical, charming, and cultured. Such sophistication made birds the key objects with which to contemplate the human estate and the morality of human behaviour.

The peacefulness of the natural world, stemming from its quietness, was especially significant in what we now see as the interwar period, when moods of pacifism expressed in the League of Nations and then in the Peace Ballot reflected the needs of society. Quietness promoted rest and reflection and yet inevitably it was alive with the small sounds of nature continuing, which lifted it away from the emptiness of silence. Most of all, quietness defined a pastoral atmosphere, a rural ideal, that countered the noisiness and disorder of the modern realm. The quietude that oozed from the smallest of green spaces, even a backyard window box, marked a rational present, free from intrusion. The quietude that radiated from the rural landscape provided a feeling for the future, secure and permanent. If there was a rhythm to British nature, it was gentle and orderly.

During this period bounded by war, nature's soundings provided continuity and stability, but listening did not remain the same. Experience on the battlefront and on the home front of WWI altered the perception of sound and its interpretation. The ears were trained to understand danger but also to seek out the sounds of safety and relief. The meanings of sounds did not necessarily change but they were accentuated. For instance, the peacefulness of the countryside could be instrumentalized by medical and military authorities because it had long been associated with recuperation. This reaching back to go forward and progress was seen too in the mythology of the nightingale's song, which was employed on the radio throughout the period of this study and then again on film for cinema audiences when its performance was in defiance of German bombs. Modern methods were put to work with old materials. The old and the new worked together when the ancient sound of the nightingale was mechanized and mediated so that all could hear it and yet still the song retained its mystique. The appeal of the natural world outside of the towns and cities was never greater than in the 1920s and 30s when so many found weekend pleasure in rambling, cycling, camping, youth hostelling, and motoring. Now that the countryside was more accessible its seductions were rediscovered, for to go there was a chance to immerse the body in sensations that towns and cities could not offer.

After WWII, music and voices were increasingly packaged by media affordable enough to access daily at home. The impresario of wildlife sounds, Ludwig Koch, continued to broaden public interest in the natural world on BBC radio, his distinctive Germanic voice helping him to become a personality in his own right; almost a David Attenborough of the 1940s. Koch's collection of hundreds of wildlife recordings was the foundation of the BBC's Natural History Unit sound library, an enterprise that in the 1950s emphasized the cultures of observation and science to British listeners and then to a new television audience. The idiosyncratic and sometimes sentimental style of Koch was superseded by a more systematic approach to communicating the natural world, which included field recording to create scientific evidence. This is not to say that popular passions for listening to and observing nature in the field diminished. The vivacity of birdlife remained central to the idea of British nature's charm and a culture of 'bird-watching'—aided by a boom in the publication of field guides and the affordability of binoculars—became a legitimate hobby for men and boys. The natural history of Britain was of widening public concern.

But I want to go back to the beginning. Because the kind of listening impressions recorded by the writer and poet Edward Thomas just before WWI seem to me as resonant and relevant as ever. Thomas wrote of winter in Suffolk:

> There are three sounds in the wood this morning—the sound of the waves that has not died away since the sea carried off church and cottage and cliff and the other half of what was once an inland wood; the sound of trees, a multitudinous frenzied sound, of rustling dead oak-leaves still on the bough, of others tripping along the path like mice, or winding up in sudden spirals and falling again, of dead boughs grating and grinding, of pliant young branches lashing, of finest twigs and fir needles sighing, of leaf and branch and trunk booming like one; and through these sounds, the song of a thrush.[2]

This conversation of sounds from the natural world, of the living and the dead, of non-human agencies jostling with the impingements of the human, abounds with potential. It is a dialogue of all earthy materials acting together. It is a dialogue that has not lost its urgency to human ears.

Acknowledgements

This, my first book, would never have materialized without the help and guidance of many along the way. First mention must go to Professors David Hendy and Michael Bull at the University of Sussex, for they were the ones who encouraged my thinking when this work was forming as a PhD some years ago. The School of Media, Film and Music at Sussex has offered a provocative and friendly environment in which to cultivate the ideas in this book. I am indebted to the Arts and Humanities Research Council for providing the funding scholarship that allowed me to devote myself to this work at Sussex. Others, including Kate Lacey and James Mansell in the UK and Karin Bijsterveld and Joeri Bruyninckx in Holland, offered invaluable support and critique of early chapters and drafts. I spent a great deal of time in the Humanities 2 reading room at the British Library in London reading, writing, and staring into space—this institution and the team there are wonderful and precious. Many other librarians and archivists helped me locate historical material over the years at the BBC's Written Archive, Imperial War Museum, British Red Cross, Royal College of Music, National Archives, and Enham Trust. The acute observations from several anonymous reviewers, working with the team at Oxford University Press, provided vital confidence. Many thanks. And my friends Tim Saunders and Lucy Rogers pored over the manuscript and the images with sustaining and cheerful vigour. Most of all, I salute Effie Paleologou for her complete and loving support throughout such a strenuous endeavour.

Notes

Introduction

1. *Listen to Britain*, 1942, dir. Humphrey Jennings and Stewart McAllister. For brief sonic analyses of *Listen to Britain*, see James Mansell, *The Age of Noise in Britain: Hearing Modernity* (Urbana: University of Illinois Press, 2017), 178–80 and E. Anna Claydon, 'National Identity, the GPO Film Unit and their Music', in *The Projection of Britain: A History of the GPO Film Unit*, ed. Scott Anthony and James G. Mansell (Basingstoke: Palgrave Macmillan, 2011), 183.

2. Edwin Ash, *The Problem of Nervous Breakdown* (New York: Macmillan, 1920), 15–16. In the 1930s, Lord Horder's Anti-Noise League became a prominent voice of senior doctors and psychologists against urban noise, which they argued damaged health.

3. MA, May 1935, quoted in *A Gleaming Landscape: 100 Years of the Guardian Country Diary*, ed. Martin Wainwright (London: Arum, 2006), 73.

4. There is a rich and interesting literature here, for example: Douglas Kahn, *Noise, Water, Meat: A History of Sound in the Arts* (Cambridge, MA: MIT Press, 2001); Karin Bijsterveld, *Mechanical Sound: Technology, Culture and Public Problems of Noise in the Twentieth Century* (Cambridge, MA: MIT Press, 2008); Hillel Schwartz, *Making Noise: From Babel to the Big Bang and Beyond* (Brooklyn: Zone, 2011); Florence Feiereisen and Alexandra Merley Hill, *Germany in the Loud Twentieth Century: An Introduction* (New York: Oxford University Press, 2012); Michael Goddard, Benjamin Halligan and Paul Hegarty, *Reverberations: The Philosophy, Aesthetics and Politics of Noise* (London: Continuum, 2012); Greg Hainge, *Noise Matters: Towards an Ontology of Noise* (New York: Bloomsbury, 2013); Marie Thompson, *Beyond Unwanted Sound: Noise, Affect and Aesthetic Moralism* (London: Bloomsbury, 2017); and Anna Snaith, *Sound and Literature* (Cambridge: Cambridge University, 2020).

5. On the social hierarchies implied by sound-making of all sorts, see, for example, Peter Denney, Bruce Buchan, David Ellison, and Karen Crawley, eds., *Sound, Space and Civility in the British World, 1700–1850* (Abingdon: Routledge, 2020) and Peter Bailey, 'Breaking the Sound Barrier: A Historian Listens to Noise', *Body and Society*, 2 (1996), 49–66.

6. Mansell, *The Age of Noise in Britain*, 25–36.

7. Emily Thompson, *The Soundscape of Modernity: Architectural Acoustics and the Culture of Listening in America, 1900–1933* (Cambridge, MA: MIT Press, 2004).

8. Anthony Jackson, 'Sound and Ritual', *Man*, 3 (1968), 293–9. Jackson's anthropological study of musical, non-musical, and non-speech sounds in non-Western cultures poses this central question.

9. In a classic study, the reception of new musical forms that challenged social norms is explored in Jacques Attali, *Noise: The Political Economy of Music* (Minneapolis: University of Minnesota Press, 1985). The BBC's early broadcasting efforts exposed diverse views about what listeners liked and disliked, see Paddy Scannell and David Cardiff, *A Social History of British Broadcasting: 1922–1939, Serving the Nation* (Oxford: Basil Blackwell, 1991).

10. Peter Cusack, 'The Favourite Sounds Project', in *Anthropology and Beauty: From Aesthetics to Creativity*, ed. Stephanie Bunn (Abingdon: Routledge, 2018), 131–47.

11. Quoted in Richard Overy, *The Morbid Age: Britain Between the Wars* (London: Allen Lane, 2009), 2. The rupturing of British life by WWI and its long-lasting consequences has been considered by many scholars including Samuel Hynes, *A War Imagined: The First World War and English Culture* (London: Bodley Head, 1990); Jay Winter, Geoffrey Parker, Mary Habeck, *The Great War and the Twentieth Century* (New Haven: Yale University Press, 2000); and David Reynolds, *The Long Shadow: The Great War and the Twentieth Century* (London: Simon and Schuster, 2014).

12. Overy, *The Morbid Age*; Hynes, *A War Imagined*, 312–14.

13. Siân Nicholas, *The Echo of War: Home Front Propaganda and the Wartime BBC, 1939–45* (Manchester: Manchester University Press, 1996), 12.

14. Raymond Williams, *Problems in Materialism and Culture* (London: Verso, 1980), 77.

15. Raymond Williams, *Keywords: A Vocabulary of Culture and Society* (London: Fontana Paperbacks, 1990), 220.

16. Kirstin Bluemel and Michael McCluskey, *Rural Modernity in Britain: A Critical Intervention* (Edinburgh: Edinburgh University Press, 2018), 2.

17. Peter Mandler, 'Against "Englishness": English Culture and the Limits to Rural Nostalgia, 1850–1940', *Transactions of the Royal Historical Society*, 7 (1997), 155–75.

18. Alexandra Harris, *Romantic Moderns: English Writers, Artists and the Imagination from Virginia Woolf to John Piper* (London: Thames & Hudson, 2010), 10–12.

19. Svetlana Boym, *The Future of Nostalgia* (New York: Basic Books, 2001), xv–xvii.

20. Quoted in Ben Shephard, *A War of Nerves: Soldiers and Psychiatrists, 1914–1994* (London: Jonathan Cape, 2000), 144.

21. Henry Williamson, *Tarka the Otter* (Harmondsworth: Penguin, 1961 [1927]), 7.

22. Alun Howkins, *The Death of Rural England: A Social History of the Countryside since 1900* (London: Routledge, 2003), 88–9 and Jeremy Burchardt, *Paradise Lost: Rural Idyll and Social Change Since 1800* (London: I. B. Tauris, 2002), 143.

23. Dennis Hardy and Colin Ward, *Arcadia for All: The Legacy of a Makeshift Landscape* (London: Mansell, 1984).

24. E. M. Nicholson and Ludwig Koch, *Songs of Wild Birds* (London: Witherby, 1951 [1936]), 183.

25. David Matless's important and influential book *Landscape and Englishness* (London: Reaktion, 1998) may have entrenched recent attention here. More recently, others have continued to concentrate on the meanings of the land, for example Paul Readman, *Storied Ground: Landscape and the Shaping of English National Identity* (Cambridge: Cambridge University Press, 2018).

26. Quoted in Keith Thomas, *Man and the Natural World* (Oxford: Oxford University Press, 1996), 265.

27. Robert Colls, *Identity of England* (Oxford: Oxford University Press, 2002), 203–5. The south country was a term that entered circulation with Edward Thomas' book of that name published in 1909.

28. David Abram, *The Spell of the Sensuous: Perception and Language in a More-Than-Human World* (New York: Vintage, 1997), 81.

29. Mark M. Smith, 'Sound—So What?', *The Public Historian*, 37 (2015), 133–4. Smith argues that the aim of aural history is to develop 'new storylines' about the experience of modernity.

30. See, for example, Mark M. Smith, *The Smell of Battle, the Taste of Siege: A Sensory History of the Civil War* (New York: Oxford University Press, 2015); Nicholas J. Saunders and Paul Cornish, eds., *Modern Conflict and the Senses* (London, Routledge, 2017); and Shane Butler and Sarah Nooter, eds., *Sound and the Ancient Senses* (New York: Routledge, 2018).

31. Maurice Merleau-Ponty, *Phenomenology of Perception*, trans. C. Smith (London: Routledge, 1992), 81–2. On the body as a sensing whole, see also Tim Ingold, 'Stop, Look, and Listen! Vision, Hearing, and Human Movement', in *The Perception of the Environment: Essays on Livelihood, Dwelling, and Skill* (London: Routledge, 2000), 243–87.

32. Michel Serres, *The Five Senses: A Philosophy of Mingled Bodies*, trans. M. Sankey and P. Cowley (London: Bloomsbury, 2008).

33. Kate Lacey, *Listening Publics: The Politics and Experience of Listening in the Media Age* (Cambridge: Polity Press, 2013).

34. On acoustic transparency of media via fidelity and listener complicity, see Jonathan Sterne, *The Audible Past: The Cultural Origins of Sound Reproduction* (Durham, NC: Duke University Press, 2003) and, in terms of radio history, Susan Douglas, *Listening in: Radio and the American Imagination* (Minneapolis: University of Minnesota, 2004), 26–30.

35. Tim Edensor, 'Thinking about Rhythm and Space', in *Geographies of Rhythm: Nature, Place, Mobilities and Bodies*, ed. Tim Edensor (Farnham: Ashgate, 2010), 7–11.

36. A. J. P. Taylor notes that the word 'England' was used in the period of this study to cover the regions of England and Wales, Great Britain, the United Kingdom, and even the British Empire: *English History 1914–45* (Oxford: Oxford University Press, 1965), v–vi.

37. David Feldman, 'Nationality and Ethnicity', in *Twentieth Century Britain: Economic, Social and Cultural Change*, ed. Paul Johnson (Harlow: Longman, 1994), 130–3 and 142–5.

38. Alex Potts, ' "Constable Country" Between the Wars', in *Patriotism: The Making and Unmaking of British National Identity III: National Fictions*, ed. Raphael Samuel (London: Routledge, 1989), 162, 182; Jeremy Burchardt, 'Rurality, Modernity and National Identity between the Wars', *Rural History*, 21 (2010), 143–50; and Mandler, 'Against "Englishness" ', 157.

39. Jeffrey Richards, *Films and British National Identity: From Dickens to Dad's Army* (Manchester: Manchester University Press, 1997), 97. Potts, ' "Constable Country" ', 175. The libertarian ideals of Edward Carpenter and other nature-worshipers in the Victorian and Edwardian periods are not typical of the pragmatic everyday concerns of most British men and women, though they are usefully examined in Jan Marsh, *Back to the Land: The Pastoral Impulse in England from 1880–1914* (London: Quartet, 1982).

40. R. Murray Schafer, *Soundscape: Our Sonic Environment and the Tuning of the World* (Rochester: Destiny Books, 1994), 8–9.

41. E. P. Thompson, 'History From Below', *Times Literary Supplement*, 7 April, 1966, 279–80. Thompson's valorization of the 'raw material of experience' as central to the construction of social history remains key for me: *The Making of the English Working Class* (London: Victor Gollancz, 1963), 9, 12. For a thoughtful examination of the place of experience in the writing of sensory history, see William Tullett, *Smell in Eighteenth-Century England: A Social Sense* (Oxford: Oxford University Press, 2019), 12–13.

42. This example comes from Potts, ' "Constable Country," ' 164–5.

43. Mark M. Smith, *How Race Is Made: Slavery, Segregation, and the Senses* (Chapel Hill: University of North Carolina Press, 2006) and Jennifer Lynn Stoever, *The Sonic Color Line: Race and the Cultural Politics of Listening* (New York: New York University Press, 2016).

44. Paul Fussell, *The Great War and Modern Memory* (New York: Oxford University Press), 251–92.

45. 'February Afternoon', in *Edward Thomas: Collected Poems and War Diary, 1917*, ed. R. George Thomas (London: Faber & Faber, 2004), 99.

Chapter 1

1. Richard Kearton, *At Home with Wild Nature* (London: Cassell, 1922), 1.

2. Erich Maria Remarque, *All Quiet on the Western Front* (London: Vintage, 1996 [1929]), 147.

3. Ian V. Hogg, *Barrage: The Guns in Action* (London: Macdonald, 1971), 26. See also Greg Goodale, *Sonic Persuasion: Reading Sound in the Recorded Age* (Urbana: University of Illinois Press, 2011), 106–31, 173–7 and Steve Goodman, *Sonic Warfare: Sound, Affect, and the Ecology of Fear* (Cambridge, MA: MIT Press, 2010), 15–26, 205–9.

4. John Lewis-Stempel's recent account of animals and plants in the lives of soldiers argues that this relationship was key to endurance at the Front. He briefly looks at the resilience and beauty of birds before moving on to horses, pets, gardening, and hunting: *Where the Poppies Blow: The British Soldier, Nature, the Great War* (London: Weidenfeld & Nicolson, 2016).

5. Music made by soldiers and for them is discussed in Emma Hanna, *Sounds of War: Music in the British Armed Forces during the Great War* (Cambridge: Cambridge University Press, 2020). A recent account of combat experience in Iraq devotes significant

attention to listening to music: see J. Martin Daughtry, *Listening to War: Sound, Music, Trauma and Survival in Wartime Iraq* (New York: Oxford University Press, 2015).

6. Paul Fussell, *The Great War and Modern Memory* (New York: Oxford University Press), 235.

7. Samuel Hynes reminds us that nearly all of the millions who fought died silently or survived, 'but in either case left no record, because they were too poor, inarticulate, unlettered or shy; or because it simply did not occur to them to write down what had happened to them': *The Soldiers' Tale: Bearing Witness to Modern War* (New York: Penguin, 1998), 32. John Laffin argues that 'the rank and file of the Great War, especially in the British and French armies, were ill-educated and often only functionally literate', *Letters from the Front, 1914–18* (London: J. M. Dent, 1973), 3. Fiona Reid has found that despite the huge volumes of personal writings from the war, there are very few accounts from ordinary shell-shocked soldiers: *Broken Men: Shell Shock, Treatment and Recovery in Britain: 1914–1930* (London: Continuum, 2010), 5.

8. Laffin, *Letters*, 2–5.

9. A. D. Gillespie, *Letters from Flanders* (London: Smith, Elder, 1916); Lawrence Housman, *War Letters of Fallen Englishmen* (New York: E. P. Dutton, 1930); Laffin, *Letters*.

10. The use of these collections minimized the selection bias that might have occurred in targeting specialist sources. I do bias my attention to the lower ranks where possible, as these voices have received less attention.

11. R. George Thomas, ed., *Edward Thomas: Collected Poems* (London: Faber & Faber, 2004), 139–72; Arthur Graeme West, *The Diary of a Dead Officer, Being the Posthumous Papers of Arthur Graeme West* (London: Allen and Unwin, 1919); Edwin Campion Vaughan, *Some Desperate Glory: The Diary of a Young Officer 1917* (Barnsley: Pen & Sword, 2010 [1981]); Wilfred Kerr, *Shrieks and Crashes: Being Memories of Canada's Corps, 1917* (Toronto: H. Rose, 1929). These have been selected for their reputation as particularly emotional portraits, having been highlighted by scholars such as Santanu Das as enabling the recovery of the 'sensuous' world of the trenches. See Santanu Das, *Touch and Intimacy in First World War Literature* (Cambridge: Cambridge University Press, 2005).

12. Edmund Blunden, *Undertones of War* (London: Penguin, 2000 [1928]); Hugh Gladstone, *Birds and the War* (London: Skeffington, 1919).

13. Eric J. Leed, *No Man's Land: Combat and Identity in World War I* (Cambridge: Cambridge University Press, 1979), 115–16, 128–9, 134. Working with first-person accounts from the main belligerent parties of the war, Leed draws out the psychological and emotion states that the war provoked and looks at how they persisted in civilian life.

14. The title of this section is taken from Isaac Rosenberg, 'Dead Man's Dump', in *Poetry of the First World War: An Anthology*, ed. Tim Kendall (Oxford: Oxford University Press, 2013), 141.

15. Alexander Watson, *Enduring the Great War: Combat, Morale and Collapse in the German and British Armies, 1914–1918* (Cambridge: Cambridge University Press,

2008), 32–3; John Terraine, *White Heat: The New Warfare 1914–18* (London: Sidgwick and Jackson, 1982), 95.

16. Robert Pickering, letter to his brother, 1 September 1915, in Housman, *War Letters*, 214.

17. Ivar Campbell, letter, winter 1915, in Laffin, *Letters*, 10.

18. Arthur Heath, letter, 6 July 1915, in Laffin, *Letters*, 47.

19. Theodore Wilson, letter to his mother, 1 March 1916, in Housman, *War Letters*, 295.

20. H. H. Munro, *The Square Egg and Other Sketches* (London: John Lane The Bodley Head, 1924), 119.

21. William Dyson, letter to his brother, 5 July 1916, in Housman, *War Letters*, 104; Edward Thomas, diary, 7 April 1917, in Thomas, ed., *Edward Thomas*, 171. Robert Graves recalled this impression five decades on, in an interview with Leslie Smith, when he said, 'Noise never stopped for one moment—ever': 'The Great Years of Their Lives', *The Listener*, 15 July 1971, 74.

22. Ernest Nottingham, letter to a friend, in Housman, *War Letters*, 199.

23. Philip J. Haythornthwaite, *The World War One Source Book* (London: Arms & Amour Press, 1994), 82–90.

24. Ernst Jünger, *Storm of Steel* (London: Penguin, 2004 [1920]).

25. Ford Madox Ford was a particularly careful listener. In two letters to Joseph Conrad in 1916 he described not just the deafening sounds he had witnessed, but how the sound of artillery fire depended on the terrain. He differentiated between wooded country, marshland, downland, or clay, as well as the effect of weather. See Tom Vandevelde, '"Are you Going to Mind the Noise?": Mapping the Soundscape of Parade's End', in *Ford Madox Ford's Parade's End: The First World War, Culture, and Modernity*, ed. A. Chantler and R. Hawkes (Amsterdam: Rodopi, 2014), 58. See also Sara Haslam, ed., *Ford Madox Ford: Parade's End Volume III: A Man Could Stand Up* (Manchester: Carcanet, 2011), xxii–iv.

26. William Dyson, letter to his brother, 5 July 1916, in Housman, *War Letters*, 103.

27. Colwyn Philipps, diary, 22 November 1914, in Colwyn Philipps, *Verses* (London: Smith, Elder, 1915), 98.

28. Remarque, *All Quiet on the Western Front*, 76.

29. Leed, *No Man's Land*, 19–20, 124.

30. Yaron Jean, 'The Sonic Mindedness of the Great War: Viewing History through Auditory Lenses', in *Germany in the Loud Twentieth Century*, 53. According to Jean, airmen and sailors had their own listening experiences and techniques, which were quite distinct from ground soldiers', 55–60. David Hendy has developed the idea of sonic mindedness by looking at Robert Graves's *Goodbye to All That*. See *Noise: A Human History of Sound and Listening* (New York: Harper Collins, 2013), 272–5. See also Axel Volmar, 'In Storms of Steel: The Soundscape of World War I and Its Impact on Auditory Media Culture During the Weimar Period', in *Sounds of Modern History*, ed. Morat, 227–36 and Elizabeth Bruton and Graeme Gooday, 'Listening in Combat: Surveillance Technologies Beyond the Visual in the First World War', *History and Technology*, 32 (2016), 213–26.

31. Edmund Blunden, *Undertones of War* (London: Penguin, 2000 [1928]), 7.

32. Blunden, *Undertones of War*, 153.
33. Blunden, *Undertones of War*, 47.
34. Kerr, *Shrieks and Crashes*, 131.
35. Mary Habeck, 'Technology in the War', in *The Great War and the Twentieth Century*, ed. Jay Winter, Geoffrey Parker, and Mary Habeck (New Haven, CT: Yale University Press, 2006), 104, note 15.
36. Blunden, *Undertones of War*, 46.
37. Wilfred Owen, 'Dulce et Decorum est', in Kendall, *Poetry of the First World War*, 155.
38. Ivar Campbell, letter, winter 195, in Laffin, *Letters*, 9.
39. J. S. Williams, letter, 3 October 1915, in Laffin, *Letters*, 58. Jack Johnson was the first African American world heavyweight boxing champion and a well-known celebrity in European culture in the years before the war.
40. My research does not concur with Bruton and Gooday, who argue that 'When the characteristic signature wail of most airborne ordnance was first heard, evasive action could be taken to leap out of the way of more deadly missiles': 'Listening in Combat', 218.
41. Blunden, *Undertones of War*, 171.
42. Charles Sorley, letter, 26 August 1916, in Housman, *War Letters*, 250.
43. Frederick Harvey, 'A True Tale of the Listening Post', in Vivien Noakes, *Voices of Silence: The Alternative Book of First World War Poetry* (Stroud: The History Press, 2006), 90–1.
44. Robert Pickering, letter to his brother, 1 September 1915, in Housman, *War Letters*, 214. The food chain has been upended, with human corpses at the bottom, rats and mice in the middle, and owls at the top.
45. William Van der Kloot, 'Lawrence Bragg's Role in the Development of Sound-Ranging in World War I', *Notes and Records of The Royal Society*, 59 (2005), 273–84. By the end of 1916, enemy artillery could be located accurately to within 25 to 50 meters. German forces were also acquiring the skills to distinguish between the size and trajectory of shells: Watson, *Enduring the Great War*, 87.
46. H. Standish Ball, 'The Work of the Miner on the Western Front, 1915–18', *Transactions of the Institution of Mining and Metallurgy*, 28 (1919), 206–7.
47. Alexander Barrie, *War Underground* (London: Star, 1981), 72. For an account of German underground listening see Julia Encke, 'War Noises on the Battlefield: On Fighting Underground and Learning to Listen in the Great War', *German Historical Institute London Bulletin*, 37 (2015), 7–21.
48. Quoted in Lewis-Stempel, *Where the Poppies Blow*, 13. When Sassoon looked at his diary he realized it contained 'lists of birds and flowers, snatches of emotion and experience'.
49. W. S. Medlicott, 'Bird Notes from the Western Front (Pas-de-Calais)', *British Birds*, 12 (1919), 272. The birds noted as common were as follows: yellowhammer, wood pigeon, corncrake, partridge and quail, chaffinch, whitethroat, crested lark, skylark, green finch, linnet, starling, tree and house sparrow, magpie, hooded crow, and kestrel.
50. Quoted in Lewis-Stempel, *Where the Poppies Blow*, 33.

51. Emma Turner, *Every Garden a Bird Sanctuary* (London: Witherby, 1935), 14.

52. See David Boyd Haycock, *Paul Nash* (London: Tate Publishing, 2002), 30–3.

53. Theodore Wilson, letter to his aunt, 27 April 1916, in Housman, *War Letters*, 296.

54. William Hodgson, 'Back to Rest', quoted in Charlotte Zeepvat, *Before Action: William Noel Hodgson and the 9th Devons: A Story of the Great War* (Barnsley: Pen & Sword, 2015), 130.

55. Edmund Blunden, *Overtones of War: Poems of the First World War*, ed. Martin Taylor (London: Duckworth, 1996), 227.

56. Quoted in Vandevelde, '"Are you Going to Mind the Noise?"', 59.

57. Quoted in Lewis-Stempel, *Where the Poppies Blow*, 49.

58. Denis Barnett, letter to his mother, 29 March 1915, in Housman, *War Letters*, 39.

59. Correspondent for *Country Life*, quoted in Lewis-Stempel, *Where the Poppies Blow*, 46.

60. R. Murray Schafer, *Soundscape: Our Sonic Environment and the Tuning of the World* (Rochester: Destiny Books, 1994), 9.

61. Thomas wrote no poetry from the trenches.

62. Thomas quoted in Judy Kendall, *Edward Thomas, Birdsong and Flight* (London: Cecil Woolf, 2014), 55. Kendall has noted the rhythms of birds and gunfire in Thomas' diary-keeping (8).

63. Diary, 1 January 1917, in Thomas, ed., *Edward Thomas*, 143.

64. March 1917 diary entries, in Thomas, ed., *Edward Thomas*, 160–9.

65. 31 March 1917, in Thomas, ed., *Edward Thomas*, 169.

66. 5 March 1917, in Thomas, ed., *Edward Thomas*, 160–1. The routine is reflected by his 21 March entry: 'Now I hardly felt as if a shell could hurt.'

67. See Habeck, 'Technology in the War', 110–11. The ferocity of weaponry sounds are also described in terms of an angry nature. Lance-Corporal George Sedding wrote that the 'earth vibrates with the gusty thunder' during an assault (in Housman, *War Letters*, 232); of bullets in the falling night: 'These ricochet off with varied noises— some with a high ringing note, others with the deep and savage hum of an angry hornet' (in Housman, *War Letters*, 233). Blunden hears bullets as aggressive insects too; he writes of 'furious insect-like zips' and 'whizzing like gnats' (*Undertones of War*, 10, 19).

68. 20 March 1917, in Thomas, ed., *Edward Thomas*, 165.

69. 16 March 1917, in Thomas, ed., *Edward Thomas*, 164.

70. 5 April 1917, in Thomas, ed., *Edward Thomas*, 170.

71. 4 April 1917, in Thomas, ed., *Edward Thomas*, 170.

72. Thomas, ed., *Edward Thomas*, 139.

73. 27 March 1917, in Thomas, ed., *Edward Thomas*, 168.

74. Thomas, ed., *Edward Thomas*, 139.

75. See Mary Douglas, *Purity and Danger: An Analysis of Concept of Pollution and Taboo* (London: Routledge, 2002), 44–5. Douglas's anthropological thinking about dirt as matter out of place has been employed by sound studies scholars to frame noise as sound out of place. I have explored elsewhere the how the disorder of trench experience could in part be managed or even reversed when the sounds of

nature were brought to auditory attention: Michael Guida, 'Nature's Sonic Order on the Western Front', *Transposition. Musique et Sciences Sociales*, special issue 'Sound, Music, Violence', ed. Luis Velasco-Pufleau (online, 2020), accessed 14 May 2020, https://doi.org/10.4000/transposition.4770.

76. See Leed, *No Man's Land*, x.

77. Watson, *Enduring the Great War*, 27–34; Michael Roper, *The Secret Battle: Emotional Survival in the Great War* (Manchester: Manchester University Press, 2009), 261.

78. Richard Donaldson, letter to his mother, 14 November 1917, in Housman, *War Letters*, 85.

79. Theodore Wilson, letter to his aunt, 27 April 1916, in Housman, *War Letters*, 298.

80. Donaldson, letter to his mother, 14 November 1917, in Housman, *War Letters*, 85.

81. Ivar Campbell, letter, date unknown, in Laffin, *Letters*, 37.

82. 24 May 1915, *Letters from Flanders*, 168.

83. Harold Chapin, letter to his wife, 4 May 1915, in Housman, *War Letters*, 71.

84. Blunden, *Undertones of War*, 157.

85. Blunden, *Undertones of War*, 178.

86. Blunden, *Undertones of War*, 157.

87. Blunden, *Undertones of War*, 93.

88. Alexander Gillespie, letter, 10 March 1915, *Letters from Flanders*, 42.

89. On the popularity of egg-collecting see Stephen Moss, *A Bird in the Bush: A Social History of Birdwatching* (London: Aurum, 2004), 93.

90. Ernest Nottingham, letter to a friend, 27 March 1916, in Housman, *War Letters*, 202.

91. Quoted in Lewis-Stempel, *Where the Poppies Blow*, 51.

92. William Dyson, letter to his brother, 5 July 1916, in Housman, *War Letters*, 103.

93. Blunden, *Undertones of War*, 24.

94. Charles Raven, *In Praise of Birds* (London: Martin Hopkinson, 1925), 5.

95. Raven, *In Praise of Birds*, 6.

96. Alexander Gillespie's letters home told of sparrows nesting precariously near his trench in May and June 1915 (*Letters from Flanders*, 143, 199). Larks nest in shell-holes in H. H. Munro's essay, 'Birds on the Western Front' (*The Square Egg*, 133–4). Monitoring the habits and migratory motions of birds in Germany during WWII provided a stabilizing routine for prisoners of war, Derek Niemann has argued in *Birds in a Cage: Warburg, Germany, 1941. Four P.O.W. Birdwatchers, the Unlikely Beginnings of British Wildlife Conservation* (London: Short Books, 2013).

97. Theodore Wilson, letter to his aunt, 27 April 1916, in Housman, *War Letters*, 296–9.

98. 24 April 1915, *Letters from Flanders*, 111.

99. Robert Sterling, letter to a friend, 18 April 1915, in Housman, *War Letters*, 263.

100. Edward Grey, *The Charm of Birds* (London: Hodder and Stoughton, 1927), 234.

101. Grey, *The Charm of Birds*, 236.

102. 14 March 1915, *Letters from Flanders*, 48.

103. Juliet Gardiner, *The Animals' War: Animals in Wartime from the First World War to the Present Day* (London: Portrait, 2006), 98–101.

104. Gladstone, *Birds and the War*, 22.

105. Push Pick, 'V. C. Canaries', *Daily Mail*, 12 September 1918, 2. *Bird Notes and News* was published by the Royal Society for the Protection of Birds.

106. Quoted in Habeck, 'Technology in the War', 102.

107. Eric Wilkinson, 'To a Choir of Birds', in Noakes, *Voices of Silence*, 83.

108. Harold Macmillan, letter, 23–4 June 1916, in Mike Webb, *From Downing Street to the Trenches: First Hand Accounts, from the Great War, 1914–1916* (Oxford: Bodleian, 2014), 191.

109. Stuart Cloete, *A Victorian Son: An Autobiography, 1897–1922* (London: Collins, 1972), 182. See also Gillespie, who hears the birds sing along with rifle fire, 11 March 1915, *Letters from Flanders*, 43. Gillespie had his own trench garden, which included a clump of violets retrieved from a flooded trench and planted in half a shrapnel shell case (83).

110. Albert Pam, 'Bird Life on the Battlefields', *The Avicultural Magazine*, 1917, 239.

111. R. Hamilton Scott, 'Birds in and Around the Firing Line', *The Avicultural Magazine*, 1918, 247–8.

112. Munro, *The Square Egg*, 133.

113. Paul Nash, letter to his wife, 7 March 1917. Tate Archive: Letters and Papers of Paul Nash. Reference TGA 8313. Accessed 14 May 2020, http://www.tate.org.uk/art/arch ive/tga-8313/letters-and-papers-of-paul-nash. See also Nash's painting, 'Spring in the Trenches', almost as ironic, if not as bitter, as 'We are Making a New World.'

114. Haslam, ed., *Ford Madox Ford*, xxiv.

115. Gladstone, *Birds and the War*, 123. Caroline Dakers reports a similar frustration of a Welsh soldier recorded by Ivor Gurney: ' "Listen to the damned bird," he said. "All through that bombardment in the pauses I could hear the infernal silly 'Cuckoo, cuckoo' sounding while Owen was lying in my arms covered with blood. How shall I ever listen again." ' *The Countryside at War* (London: Constable, 1987), 173.

116. The origins of the 'Keep Calm and Carry On' slogan have been located in WWII with the activity of the Ministry of Information, but the roots of this idea are ear- lier. See Lynda Mugglestone, 'Rethinking the Birth of an Expression. Keeping Calm and "Carrying On" in World War One', *English Words in Wartime*, 2 August 2016, accessed 14 May 2020, https://wordsinwartime.wordpress.com/2016/08/02/ret hinking-the-birth-of-an-expression-keeping-calm-and-carrying-on-in-world- war-one/.

117. Gladstone, *Birds and the War*, 111–14.

118. Gladstone, *Birds and the War*, 122.

119. Lewis-Stempel, *Where the Poppies Blow*, 48.

120. Gladstone, *Birds and the War*, 104.

121. Nature myths were strong in German culture and war experience too. For instance, Erich Maria Remarque's protagonist finds himself transported by the wind in the leaves of poplar trees to his boyhood: *All Quiet on the Western Front*, 85–8. Alex Potts argues that every bit as potent as the English meadow and village was the German forest (Wald) and village (Dorf): ' "Constable Country," ' 162. George Mosse argues that the symbolism of the tree and the wood was specifically German and associ- ated with 'innocent nature'. To be buried within 'heroes' groves' of trees was to be

given a restful memorial: *Fallen Soldiers: Reshaping the Memory of the World Wars* (New York: Oxford University Press, 1991), 110.

122. 'Bird's Song on Battlefield', *Daily Mail*, 27 April 1915, 3.

123. The idea of the brutal 'Hun' in modern German history appeared in 1900 when the term was employed by Kaiser Wilhelm II to evoke the ruthlessness he expected from his troops in the suppression of the boxer Rebellion in China. The British developed the stereotype of the German Hun through propaganda about their apparent atrocities, using it as the basis for a moral offensive against a German society said to be founded upon militaristic values. This propaganda also made concrete to British combatants and civilians the unimaginable consequences of defeat. See David Welch, *Germany, Propaganda and Total War, 1914–1918* (New Brunswick: Rutgers University Press, 2000), 61–2. Nicoletta F. Gullace has explored the visual depictions of the Hun in 'Barbaric Anti-Modernism: Representations of the "Hun" in Britain, North America, Australia, and Beyond', in *Picture This: World War I Posters and Visual Culture*, ed. Pearl James (Lincoln: University of Nebraska Press, 2009), 61–78.

124. Mark Cocker and Richard Mabey, *Birds Britannica* (London: Chatto & Windus, 2005), 442–3.

125. Gladstone, *Birds and the War*, 25.

126. 'Soldier and Song Bird', *Daily Mail*, 21 January 1916, 3.

127. Samuel Foster Damon, *A Blake Dictionary: The Ideas and Symbols of William Blake* (Hanover: University Press of New England, 1988), 234; Percy Bysshe Shelley, 'To A Skylark', in *The Bird-Lovers' Anthology*, ed. C. Scollard and J. B. Rittenhouse (Boston: Houghton Mifflin, 1930), 56. Lark symbolism endures in literature and manages to just avoid cliché in Sebastian Faulk's novel *Birdsong* (London: Vintage, 1993), where a lark 'singing in the unharmed air' signals the end, not of a bombardment or battle, but of the war, for protagonist Stephen Wraysford (485).

128. Lewis-Stempel, *Where the Poppies Blow*, 51.

129. Quoted in Lewis-Stempel, *Where the Poppies Blow*, 79–80.

130. Fussell, *The Great War*, 51–2. Siegfried Sassoon recalled thinking that 'the sky was one of the redeeming features of the war', Fussell points out.

131. Leed discusses very interestingly the connections between immobility in the trenches and war neurosis: *No Man's Land*, 180–92.

132. D. Forsythe, 'Functional Nerve Disease and the Shock of Battle', *The Lancet*, 25 December (1915): 1400.

133. The idea of the strangeness of trench existence is discussed by Hynes, *The Soldiers' Tale*, 18, 19, 52, 53.

134. Rowland Feilding, letter, 8 October 1917, in Laffin, *Letters*, 95.

135. Blunden, *Undertones of War*, 13.

136. David Jones, 'In Parenthesis', in Kendall, *Poetry of the First World War*, 202–3.

137. Charles Sorley, letter, 26 August 1915, in Housman, *War Letters*, 251.

138. Cecil Day-Lewis, ed., *The Collected Poems of Wilfred Owen* (New York: New Directions Books, 1965), 22. 'He plunges at me guttering, choking, drowning' comes after the gas attack in 'Anthem for Doomed Youth': Kendall, *Poetry of the First World War*, 155.

139. 'In our mess we never allow any mention of anything depressing', Captain Colwyn Philipps wrote to his mother, 13 November 1914, in Housman, *War Letters*, 213.

140. John Keegan, *The Face of Battle* (New York: Penguin, 1978), 225. John Laffin's collection of letters from the Front makes clear the importance of the love felt between fellow men: *Letters*, 9, 34, 108.

141. Earnest Nottingham, letter, in Housman, *War Letters*, 202.

142. Isaac Rosenberg, 'Returning we Hear the Larks', in Kendall, *Poetry of the First World War*, 139.

143. Fussell, *The Great War*, 242.

144. John W. Streets, 'Shelley in the Trenches', in Noakes, *Voices of Silence*, 83.

145. Leed, *No Man's Land*, 22. Burial alive was not uncommon and instances could often precipitate psychological breakdown.

146. Philipps, *Verses*, 126.

147. Frederick Keeling, letter, 25 March 1916, in Housman, *War Letters*, 164. There are other examples of birdsong triggering nostalgia. Lieutenant Trotter in R. C. Sherriff's autobiographical play 'Journey's End' comments: 'Funny about that bird. Made me feel quite braced up. Sort of made me think about my garden of an evening—walking round in me slippers after supper, smoking me pipe': Lewis-Stempel, *Where the Poppies Blow*, 32.

148. Ivor Gurney, 'Birds', in *The Poetry of Birds*, ed. Simon Armitage and Tim Dee (London: Penguin, 2011), 54.

149. Paul Nash, letter to his wife, 26 April 1917. Tate Archive: Letters and Papers of Paul Nash. Reference TGA 8313.

150. Ralph Barker, *The Royal Flying Corps in World War One* (London: Robinson, 2002), 278. See Mosse on German pilots as chivalric beings and their status as emblematic of a new Germany before and during the war: *Fallen Soldiers*, 118–121.

151. Theodore Wilson, letter to his aunt, 27 April 1916, in Housman, *War Letters*, 297.

152. Moss, *A Bird in the Bush*, 116.

153. 14 February 1917, in Thomas, ed., *Edward Thomas*, 155.

154. Leed, *No Man's Land*, 134.

155. John Gray has touched on these ideas in his analysis of human-animal relations, considering in particular J. A. Baker's *The Peregrine*. John Gray, *The Silence of Animals: On Progress and Other Modern Myths* (London: Penguin, 2014), 148–9.

156. Remarque, *All Quiet on the Western Front*, 101.

157. Siegfried Sassoon, 'Everyone Sang', in Armitage and Dee, *The Poetry of Birds*, 177–8.

158. Tammy M. Proctor, 'The Everyday as Involved in War', in *International Encyclopaedia of the First World War*, ed. Ute Daniel et al. (Berlin: Freie Universität Berlin, 2014), accessed 14 May 2020, http://dx.doi.org/10.15463/ie1418.10453.

Chapter 2

1. The first verse of a poem by an officer being treated for shell-shock at Craiglockhart Hospital near Edinburgh, who signed himself North British: 'Be Still, My Soul', *The*

Hydra, 21 July 1917, 9. The First Word War Poetry Digital Archive, accessed 20 July 2016, http://www.oucs.ox.ac.uk/ww1lit/collections/document/3133/1912.

2. Robert Graves wrote of Sassoon's wishes based on a letter he had received from him: *Goodbye to All That* (London: Penguin, 1960 [1929]), 211–12.

3. Fussell, *The Great War*, 231.

4. The personal and societal shock of the war has been explored by many scholars, though not in relation to sound and listening. For example: Gerard J. DeGroot, *Blighty: British Society in the Era of the Great War* (Harlow: Pearson, 1996); Jay Winter, *The Great War and the British People* (London: Palgrave Macmillan, 2003); and Arthur Marwick, *The Deluge: British Society and the First World War* (London: Palgrave Macmillan, 2006).

5. Peter Leese, *Shell Shock: Traumatic Neurosis and the British Soldiers of the First World War* (Basingstoke: Palgrave Macmillan, 2002). Pat Barker's widely read *Regeneration*, and the subsequent film, about psychiatrist W. H. R. Rivers and his officer patients Siegfried Sassoon, Wilfred Owen, and the fictional Billy Prior, has contributed to the continued cultural resonance of shell shock, as have the many commemorative portrayals of the war at its centenary.

6. Jay Winter, 'Shell Shock', in *The Cambridge History of the First World War*, ed. Jay Winter (Cambridge: Cambridge University Press, 2014), 330. See also Edgar Jones and Simon Wessely, 'Psychiatric Battle Casualties: An Intra- and Interwar Comparison', *British Journal of Psychiatry*, 178 (2001), 242–7.

7. Anthony Richards, 'The British Response to Shell-Shock: An Historical Essay', in *Report of the War Office Committee of Enquiry into 'Shell-Shock'* (London: Imperial War Museum Military Handbook Series, 2004), iv.

8. Hynes, *A War Imagined*, 465–9. Overy, *The Morbid Age*.

9. The historiography of shell shock is vast, though texts about the treatment of the condition may be fewer than the analyses influenced by the work of Michel Foucault, Andrew Scull, and Thomas Szasz of social control exercised by doctors and political interests. For a brief review of this latter kind of scholarship, see Mark O. Humphries and Kellen Kurchinski, 'Rest, Relax and Get Well: A Re-Conceptualisation of Great War Shell Shock Treatment', *War & Society*, 27 (2008): 90–1. Key texts about the treatment and cultural impact of shell shock that I have referred to are Elaine Showalter, *The Female Malady: Women, Madness and English Culture, 1830–1980* (New York: Virago, 1987); Ted Bogacz, 'War Neurosis and Cultural Change in England, 1914–22: The Work of the War Office Committee of Enquiry into "Shell-Shock,"' *Journal of Contemporary History*, 24 (1989), 227–56; Joanna Bourke, *Dismembering the Male: Men's Bodies, Britain and the Great War* (London: Reaktion, 1999); Paul Lerner, 'Psychiatry and Casualties of War in Germany, 1914–18', *Journal of Contemporary History*, 35 (2000), 13–28; Ben Shephard, *A War of Nerves: Soldiers and Psychiatrists, 1914–1994* (London: Jonathan Cape, 2000); Jay Winter, 'Shell-Shock and the Cultural History of the Great War', *Journal of Contemporary History*, 35 (2000), 7–11; Edgar Jones and Simon Wessely, *Shell Shock to PTSD: Military Psychiatry from 1900 to the Gulf War* (Hove: Psychology Press, 2005); Peter Barham, *Forgotten Lunatics of the Great War* (New Haven, CT: Yale University Press, 2007);

Paul Lerner, *Hysterical Men: War, Psychiatry, and the Politics of Trauma in Germany, 1890–1930* (Ithaca, NY: Cornell University Press, 2008); Tracey Loughran, 'Hysteria and Neurasthenia in Pre-War Medical Discourse and in Histories of Shell-Shock', *History of Psychiatry*, 19 (2008), 25–46; Fiona Reid, *Broken Men: Shell Shock, Treatment and Recovery in Britain: 1914–1930* (London: Continuum, 2010); Jason Crouthamel and Peter Leese, *Psychological Trauma and the Legacies of the First World War* (Palgrave Macmillan, 2016); and Fiona Reid, *Medicine in First World War Europe: Soldiers, Medics, Pacifists* (London: Bloomsbury, 2017).

10. Brendan Kelly, '*He Lost Himself Completely': Shell Shock and Its Treatment at Dublin's Richmond War Hospital, 1916–1919* (Dublin: Liffey Press, 2014), 61–83.

11. Reid, *Broken Men*, 75.

12. Reid, *Broken Men*, 34.

13. Winter, 'Shell Shock', 325. Winter footnotes his experience of meeting men like this at Warwick General Hospital in 1978, where they had been for some sixty years.

14. Michael Roper has made innovative use of family accounts in order to get closer to the recollections of emotional experience in communities and across generations: *The Secret Battle: Emotional Survival in the Great War* (Manchester: Manchester University Press, 2009).

15. See Mark Harrison, *The Medical War: British Military Medicine in the First World War* (Oxford: Oxford University Press, 2010), 113–16. Harrison points out that hospitals like the famous Craiglockhart near Edinburgh were by no means typical, many shell-shock patients never making it back to hospitals near base, let alone Britain.

16. See, for example, Peter Howorth, 'The Treatment of Shell Shock: Cognitive Therapy Before Its Time', *Psychiatric Bulletin*, 24 (2000), 225–7.

17. W. H. R. Rivers at Craiglockhart Hospital became well-known for his techniques, informed by Sigmund Freud's theories, for re-structuring the way a soldier viewed and interpreted past experiences. One of his wartime publications sets out his views: 'An Address on the Repression of War Experience', *The Lancet*, 2 February 1918.

18. *Report of the War Office Committee of Enquiry into 'Shell-Shock'* (London: HSMO, 1922). Southborough's report saw the condition as an excuse for a way out of combat. It asked for the term 'shell shock' to banned, accepting no pathology and no notion of complex emotional breakdown. For critiques of the report, see Barham, *Forgotten Lunatics of the Great War*, 233–8 and Bogacz, 'War Neurosis and Cultural Change in England'.

19. *Report of the War Office*, 158.

20. *Report of the War Office*, 192.

21. Jeffrey S. Reznick, *Healing the Nation: Soldiers and the Culture of Caregiving in Britain During the Great War* (Manchester: Manchester University Press, 2005), 21–7.

22. While Jay Winter emphasizes artillery noise in the development of shell shock, Joanna Bourke offers case studies that indicate the horror of face-to-face killing and seeing one's comrades mutilated were the chief causes of the phenomenon. See Winter, 'Shell Shock' and Bourke, *Dismembering the Male*.

23. D. Forsythe, 'Functional Nerve Disease and the Shock of Battle', *The Lancet*, 25 December 1915, 1400. Forsythe also reports: 'The only treatment during the acuter stage comprises three items—physical rest in bed, mental quiet, and good food.'

24. Myers first coined the term 'shell shock' in 1915: C. S. Myers, 'A Contribution to the Study of Shell Shock: Being an Account of Three Cases of Loss of Memory, Vision, Smell, and Taste, Admitted into the Duchess of Westminster's War Hospital, Le Touquet', *The Lancet*, 13 February 1915.

25. *Report of the War Office*, 124.

26. Christine Hallett, ' "This Fiendish Mode of Warfare": Nursing the Victims of Gas Poisoning in the First World War', in *One Hundred Years of Wartime Nursing Practices, 1854–1953*, ed. Jane Brooks and Christine Hallett (Manchester: Manchester University Press, 2015), 83.

27. Violetta Thurstan, *A Text Book of War Nursing* (London: GP Putnam, 1917), 139.

28. *Report of the War Office*, 160. A. G. Macdonnell recalled that those at Craiglockhart Hospital 'who disliked noise were allotted rooms on the main road' by Colonel Balfour-Graham, who replaced W. H. R. Rivers. See Thomas Webb, ' "Dottyville"— Craiglockhart War Hospital and Shell-Shock Treatment in the First World War', *Journal of the Royal Society of Medicine*, 99 (2006), 343.

29. Michael S. Kimmel and Amy Aronson, eds., *Men and Masculinities: Volume 1* (Santa Barbara: Clio, 2004), 554. See also Anne Stiles, 'The Rest Cure, 1873–1925', *Britain, Representation, and Nineteenth-Century History*, accessed 1 August 2017, http://www.branchcollective.org/?ps_articles=anne-stiles-the-rest-cure-1873-1925.

30. For example, captain Wilfred Harris recommended the rest cure in his book *Nerve Injuries and Shock* (London: Henry Frowde, 1915). However, soldiers found many treatments for battle injury laughable and often mocked medical care, including the rest cure: see Reid, *Broken Men*, 74–6 and Reznick, *Healing the Nation*, 79–82.

31. Florence Nightingale, *Notes on Nursing: What It Is and What It Is Not* (Philadelphia: Stern, 1946 [1859]), 33. Nightingale's medical insight reflected societal preferences from her own elite background for quietude, along with the religious and meditative commitments she had made in her training as a deaconess. See Hillel Schwartz, 'Inner and Outer Sancta: Earplugs and Hospitals', in *The Oxford Handbook of Sound Studies*, ed. Trevor Pinch and Karin Bijsterveld (New York: Oxford University Press, 2012), 274–80.

32. See, for example, Amy Millicent Ashdown, *A Complete System of Nursing* (London: J. M. Dent, 1917).

33. Stiles, 'The Rest Cure'.

34. George Beard, *American Nervousness: Its Causes and Consequences* (New York: G. P. Putnam's, 1881), 106.

35. For Beard, modern civilization was the cause of the problem of nervousness, especially 'steam power, the periodical press, the telegraph, the sciences, and the mental activity of women' (vi). For British perspectives, see Mansell, *The Age of Noise*, 25–37; Michael Neve, 'Public Views of Neurasthenia: Britain, 1880–1930', in *Cultures of Neurasthenia from Beard to the First World War*, ed. Marijke Gijswijt-Hofstra and Roy Porter (New York: Rodopi, 2001), 141–59.

36. On asylum environments, see, for example, Jonathan Andrews and Andrew Scull, *Undertaker of the Mind: John Munro and Mad-Doctoring in Eighteenth Century England* (Berkeley and Los Angeles: University of Californian Press, 2001), 35 and Clare Hickman, 'Cheerfulness and Tranquillity: Gardens in the Victorian Asylum,' *The Lancet*, 1 (December 2014): 506–7. For wider assessments of nature's healing potential, see Clare Hickman, *Therapeutic Landscapes: A History of English Hospital Gardens Since 1800* (Manchester: Manchester University Press, 2013); Helen Bynum, *Spitting Blood: The History of Tuberculosis* (Oxford: Oxford University Press, 2012); and John Hassan, *The Seaside, Health and the Environment in England and Wales Since 1800* (Aldershot: Ashgate, 2003).

37. Janelle Stanley, 'Inner Night and Inner Light: A Quaker Model of Pastoral Care for the Mentally Ill,' *Journal of Religion and Health*, 49 (2010), 547–59.

38. Jerry White, *Zeppelin Nights: London in the First World War* (London: Vintage, 2015), 167–8.

39. Arthur Haydon, 'Homes for Shell-Shock Cases,' *Daily Mail*, 29 March 1916, 4.

40. 'Cure by Spring Hues,' *Daily Mail*, 27 September 1917, 3.

41. Mary E. Mitchison, 'A Debt to the Wounded,' *Times*, 23 May 1916, 9.

42. 'Rest Gardens for the Wounded,' *Times*, 24 May 1916, 11.

43. White, *Zeppelin Nights*, 168.

44. John Picker, *Victorian Soundscapes* (Oxford: Oxford University Press, 2003), 66.

45. This poster in fact shows a soldier in the hodden grey kilt and glengarry bonnet of a Scottish regiment.

46. Potts, 'Constable Country,' 175. See also Colls, *Identity of England*, 203–6.

47. *Defence of the Realm Manual*, 6th ed. (London: HMSO, 1918), 104. The Defence of the Realm Act was enacted in August 1916. See also White, *Zeppelin Nights*, 168.

48. Lord Knutsford, *In Black and White* (London: Edward Arnold: 1927), 268–70. Knutsford made a successful public appeal to raise £10,000 to run the Palace Green hospital for two years: 'The Care of the Wounded,' *British Journal of Nursing*, 5 February 1916, 120.

49. Quoted in Barham, *Forgotten Lunatics*, 43.

50. Lord Knutsford, 'Lord Knutsford's Appeal,' *Times*, 13 November 1914, 9.

51. 'A Kensington Hospital for Officers,' *Times*, 18 January 1915, 4.

52. 'A Kensington Hospital for Officers.'

53. Aubrey House Auxiliary Hospital, Lost Hospitals of London, accessed 3 September 2017, http://ezitis.myzen.co.uk/aubreyhouse.html.

54. 'Shell-Shock Men,' *Daily Mail*, 8 November 1917, 5.

55. Seymour Hicks, 'Shell-Shock Cases in Air Raids,' *Daily Mail*, 27 December 1917, 2.

56. Caroline Dakers, *The Countryside at War* (London: Constable, 1987), 36.

57. Thomas Lumsden, 'Nerve-Shattered Pensioners,' *Times*, 22 August 1917, 9.

58. Lumsden, 'Nerve-Shattered Pensioners,' 8.

59. 'Country Air for Shell-Shock,' *Daily Mail*, 26 November 1917, 5; 'Disabled Soldiers In Civil Hospitals,' *Times*, 23 January 1918, 3.

60. Thomas Lumsden, letter, *British Medical Journal*, 21 January 1920, 131.

61. See Humphries and Kurchinski, 'Rest, Relax and Get Well', about the Canadian military-medical authorities' use of spas in Kent and Derbyshire.

62. Promotional brochure from 1903, History of Craiglockhart, War Poets Collection at Edinburgh Napier University, accessed 3 September 2017, http://www2.napier.ac.uk/warpoets/1800.htm#1800.

63. Private papers of Lieutenant J. H. Butlin, letters sent from Craiglockhart Hospital to Basil Burnett Hall between May and July 1917, Imperial War Museum (IWM), catalogue number 7915.

64. North British, 'Be Still, My Soul', 9. First Word War Poetry Digital Archive, accessed 20 July 2016, http://www.oucs.ox.ac.uk/ww1lit/collections/document/3133/1912. The views of shell-shocked men are few and far between. *The Hydra* provides evidence of a longing and fondness for the imagined pastoral, an escape from the hauntings of the trenches and a vision of the nation at peace, which the surroundings at Craiglockhart may well have nurtured. The considerate ways of Rivers were not the only ways at Craiglockhart. Colonel Balfour-Graham, who replaced Rivers, favoured disciplinary regimes that could involve toughening-up by exposure to noise.

65. The differences between the frequency of occurrence of shell shock, diagnosis, and treatment in officers and ranks are still contended in relation to class, culture, and medical thought. See, for example, Showalter, *The Female Malady*, 174 and Winter, 'Shell Shock', 331. Clearer is that physicians were prone to diagnose an officer's mental collapse as neurasthenia (which, before the war, was a cultivated man's affliction) and to judge another ranks' mental breakdown as hysteria, for which more punitive treatments were recommended. See Bogacz, 'War Neurosis and Cultural Change', note 71.

66. Barham, *Forgotten Lunatics*, 43–4.

67. Linda Bryder, 'Papworth Village Settlement—a Unique Experiment in the Treatment and Care of the Tuberculosis?', *Medical History*, 28 (1984), 372–90. At Papworth there were many common features with the Village Centre idea, such as attention to the whole person, treatment with fresh air and craft work, and the establishment of a permanent community where emphasis was placed on the happy family for health.

68. Frederick Milner's Eden Manor promotional booklet, The Ex-Services Welfare Society, National Archives, Kew, reference PIN 15/2499. Milner was a founder in 1919 of the Ex-Services Welfare Society and led the opening of a rehabilitative home called Eden Manor in Kent in 1924 for shell-shocked soldiers. The Society argued that Eden Manor was distinguished by peaceful and picturesque surroundings, very different to the pauper asylums they claimed had absorbed thousands of shell-shocked men. See Reid, *Broken Men*, 100–62; Barham, *Forgotten Lunatics*, 293–9.

69. For example, Frederick Milner, letter, *John O'Groat Journal*, 14 October 1921, 3.

70. Village Centres Council Annual Report, 1919, Enham Village Centre archive reference EAA/1/3/1/1/2, 18–20.

71. Warwick Draper, 'Village Centres for Cure and Training', *Recalled to Life*, 21 April 1918, 346.

72. Draper, 'Village Centres for Cure and Training', 345–6.

73. Draper, 'Village Centres for Cure and Training', 346.

74. Draper, 'Village Centres for Cure and Training', 347.

75. Enham Village Centre promotional brochure (undated), IWM archive, catalogue number K.91/2253, 3.

76. Enham Village Centre promotional brochure. In 1935 the depiction of pastoral imagery is still very prominent in the Village Centres Council Annual Report, 1935, IWM archive, catalogue number KS.84/55.

77. 'The Opening of Enham Village Centre', *British Medical Journal*, 8 November 1919, 610. Village Centres Council Annual Report, 1920, Enham Village Centre archive reference EAA/1/3/1/1/2.

78. This was part of Baldwin's speech to the 6 May 1924 meeting of the Royal Society of St George, reported in 'English Traits. Mr Baldwin's Review', *Times*, 7 May 1924, 16. See Howkins, 'The Discovery of Rural England', 105. For a discussion of the links between Conservatism and rurality, see Burchardt, *Paradise Lost*, 104–6.

79. Quoted in Howkins, 'The Discovery of Rural England', 105.

80. By April 1919 there were nearly 400,000 unemployed ex-serviceman in Britain: Jon Lawrence, 'The First World War and Its Aftermath', in *Twentieth Century Britain: Economic, Social and Cultural Change*, ed. Paul Johnson (Harlow: Longman, 1994), 161.

81. *Reveille* was published by HMSO from August 1918 to February 1919 and before that it was called *Recalled to Life* (June 1917 to April 1918) and edited by the Liberal politician and writer Lord Charnwood.

82. John Galsworthy, 'The Gist of the Matter', *Reveille*, 1 August 1918, 3. Samuel Hynes has pointed out that Galsworthy was well known as a man who could not accept the idea of a twentieth-century, urban, industrial England, in *The Edwardian Turn of Mind* (Princeton, NJ: Princeton University Press), 63.

83. Galsworthy, 'The Gist of the Matter', 4–5.

84. Galsworthy, 'Heritage: An Impression', 303.

85. Galsworthy, 'Heritage: An Impression', 303.

86. 'Village Settlements for the Disabled', *The Lancet*, 3 November 1917, 685.

87. 'Village Settlements for Disabled Ex-Servicemen', *The Lancet*, 3 November 1917, 692–3.

88. 'Village Settlements for Disabled Ex-Servicemen', 692.

89. Village Centres Council Annual Report, 1920, Enham Village Centre archive reference EAA/1/3/1/1/2. These features were reported in the medical press, for example 'The Enham Village Centre for the Re-education of Men Disabled in the War', *The Lancet*, 9 August 1919, 252.

90. Dan McKenzie, *City of Din, A Tirade Against Noise* (London: Adlard, 1916). In the 1930s McKenzie became an active commentator about the unhealthiness of city noise and promoted the benefits of rhythmic sound, which he said reduced the onset and intensity of fatigue. See Dan McKenzie, 'Noise and Health', *British Medical Journal*, 6 October 1934, 636.

91. Arthur McIvor, 'Employers, the Government, and Industrial Fatigue in Britain, 1890–1918', *British Journal of Industrial Medicine*, 44 (1987), 724–32.

92. Arthur McIvor, 'Manual Work, Technology and Industrial Health, 1918–39', *Medical History*, 31 (1987), 167–70.
93. Letter from Field Marshal Haig, inside front cover of Village Centres Council Annual Report, 1921, Enham Village Centre archive reference EAA/1/3/1/1/3.
94. Major C. Reginald Harding, 'Land Settlement and the Disabled', *Reveille*, 1 February 1919, 467.
95. Harding, 'Land Settlement and the Disabled', 471.
96. These rhythms provided the chance perhaps for men to entrain with their environment. Human synchronization with musical rhythm has been explored by Martin Clayton, Rebecca Sager, and Udo Will, 'In Time with the Music: The Concept of Entrainment and Its Significance for Ethnomusicology', *European Seminar in Ethnomusicology Counterpoint*, 1 (2005), 5–7. See also David Hendy, *Noise: A Human History of Sound and Listening* (New York: Harper Collins, 2013), 16–18.
97. 'A Rural Life for Health and Restoration', *Fruit Grower*, 29 January 1920, 187.
98. Major G. V. Stockdale, Village Centres Annual Report, 1923, 8. It is important to note that men with physical wounds were rarely given a diagnosis of neurasthenia, as if physical wounds had no other consequences, and physical and psychiatric problems could not co-exist. See Winter, 'Shell Shock', 331–2.
99. Village Centres Council Annual Report, 1935, IWM archive, catalogue number KS.84/55.
100. Hurst intended to minimize pension claims as part of his treatment plan. See A. F. Hurst and J. L. M. Symns, 'The Rapid Cure of Hysterical Symptoms in Soldiers', *The Lancet*, 6 August 1918, 140.
101. Arthur Hurst quoted in Melanie Dunn, 'Hysterical War Neuroses: A Study of Seale-Hayne Neurological Military Hospital, Newton Abbot, 1918–1919' (MA diss., University of Exeter, 2009), 18.
102. Hurst and Symns, 'The Rapid Cure of Hysterical Symptoms', 140.
103. *War Neurosis: Netley Hospital, 1917*, film, 1918 (Arthur Hurst and J. L. M. Symns). See A. F. Hurst, 'Cinematograph Demonstration of War Neuroses', *Proceedings of the Royal Society of Medicine*, 11 (1918). For a critical appraisal of Hurst's film evidence of his cures, see Wendy Holden, *Shell Shock: The Psychological Impact of War* (London: Channel 4 Books, 1998), 19.
104. The most detailed analysis of Hurst's films of his treatment successes makes no mention of these prominent scenes: Edgar Jones, 'War Neuroses and Arthur Hurst: A Pioneering Medical Film About the Treatment of Psychiatric Battle Causalities', *Journal of the History of Medicine and Allied Sciences*, 67 (2012).
105. Hurst and Symns, 'The Rapid Cure of Hysterical Symptoms', 140.
106. Denis Winter, *Death's Men: Soldiers of the Great War* (London: Penguin, 1979), 244.
107. Philip Gibbs, *Now It Can be Told* (New York: Harper and Brothers, 1920), 548.
108. J. Winter quoted in Winter, *Death's Men*, 243.
109. Graves, *Goodbye to All That*, 238.
110. Dakers, *The Countryside at War*, 112.
111. Siegfried Sassoon, 'Repression of War Experience', in Kendall, *Poetry of the First World War*, 98 and 255 note.

112. Mary Beazley, 'The Sound of Flanders Guns,' in Noakes, *Voices of Silence*, 297. Axel Volmar argues that in Germany after the war, ex-servicemen and those who had not come into direct contact with the war heard differently: 'In Storms of Steel', 240–1.

113. Lynda Mugglestone, '"That Siren Call . . . " The Diverse Language of Air-Raid Precautions in 1916', *English Words in Wartime*, 17 May 2016, accessed 5 September 2017, https://wordsinwartime.wordpress.com/2016/05/17/that-siren-call-the-dive rse-language-of-air-raid-precuations-in-1916/.

114. See White, *Zeppelin Nights*, 214–15, 252–3.

115. William Sandhurst quoted in White, *Zeppelin Nights*, 254.

116. White, *Zeppelin Nights*, 215.

117. Graves, *Goodbye to All That*, 220.

118. Dennis Hardy and Colin Ward, *Arcadia for All: The Legacy of a Makeshift Landscape* (London: Mansell, 1984). See also Gillian Darley, *Villages of Vision: A Study of Strange Utopias* (Nottingham: Five Leaves, 2007). The plotland movement coincided with 1919 legislation for homecoming soldiers, which provided funds to establish them on small-holdings or allotments: see Howkins, *The Death of Rural England*, 88–9.

119. Hardy and Ward, *Arcadia for All*, 191.

120. Hardy and Ward, *Arcadia for All*, 276.

121. Edwin Ash, *The Problem of Nervous Breakdown* (New York: Macmillan, 1920), 15–16. Though his peaceful sanatoria to rebuild national nerve strength were unaffordable on the scale he suggested, Ash's concerns reveal the perception of a brittle society in need of quiet rest.

122. Dan McKenzie, *City of Din*, 20. McKenzie has almost nothing to say about shell shock in this book published during the war. He becomes a vocal campaigner in the city noise abatement movement of the 1930s led by Lord Horder.

123. Dan McKenzie, *Aromatics of the Soul: A Study of Smells* (London: Heinemann, 1923), 145.

124. McKenzie, *Aromatics of the Soul*, 153–4.

125. George Revill, 'Music and the Politics of Sound: Nationalism, Citizenship, and Auditory Space', *Environment and Planning D: Society and Space*, 18 (2000), 599–601.

126. David Matthews, 'The Music of English Pastoral', in *Town and Country*, ed. A. Barnett and R. Scruton (London: Jonathan Cape, 1998), 84.

127. Eric Saylor, '"It's Not Lambkins Frisking at All": English Pastoral Music and the Great War', *The Musical Quarterly*, 91 (2009), 40, 54. Saylor points out that Vaughan Williams offered himself as a forty-two-year-old to be an ambulance driver and lost his young friend George Butterworth. His Pastoral Symphony was wartime music responding to his experiences, though sometimes it was misinterpreted at the time as suggestive of 'lambkins frisking'. His 'concerns were those of a modernist driven by the need to make sense of an alienating post-war world while armed only with pre-war artistic experience' (48–9).

128. Saylor, '"It's Not Lambkins Frisking at All,"' 41. This important and overdue argument is elaborated convincingly, including a close investigation of pastoral music in

wartime, in Eric Saylor, *English Pastoral Music: From Arcadia to Utopia, 1900–1955* (Urbana: University of Illinois, 2017).

129. Saylor, ' "It's Not Lambkins Frisking at All," ' 44.

130. Howkins, *The Discovery of Rural England*, 92.

131. Hynes, *A War Imagined*, 311–14. Hynes argues that talk of the end of civilization after the war was widespread and that the highest values of culture were seen to be needed to reconstruct the nation. However, he also suggests that many perceived a chasm left by the war, which meant that the resources of the old England before the war would be difficult to reach and draw upon.

132. Simon Miller, 'Urban Dreams and Rural Reality: Land and Landscape in English Culture, 1920–45', *Rural History*, 6 (1995): 89–9. Miller argues that Baldwin's 'England' speech was well known. Baldwin's rural idealism did not include the social and commercial aspects of agriculture, rather it focused on the culture, leisure, and wisdom that might be imparted from communion with the countryside and the seasons.

133. Ernest Pulbrook, *The English Countryside* (London: B. T. Batsford, 1915), 106.

134. Winter has argued that individuals turned to the past, returning to 'ordered patterns and themes', to make sense of the war: *Sites of Memory, Sites of Mourning: The Great War in European Cultural History* (Cambridge: Cambridge University Press, 1995), 6–7.

135. On silence and commemoration after war from international perspectives, see E. Ben-Ze'ev, R. Genie, and J. Winter, eds., *Shadows of War: A Social History of Silence in the Twentieth Century* (Cambridge: Cambridge University Press, 2010). The foundational text about the British Armistice is Adrian Gregory, *The Silence of Memory: Armistice Day, 1919–1946* (Oxford: Berg, 1994). On musical alternatives, see James G. Mansell, 'Musical Modernity and Contested Commemoration at the Festival of Remembrance 1923–1927', *Historical Journal*, 52 (2009), 433–54; Rachel Cowgill, 'Canonizing Remembrance: Music for Armistice Day at the BBC, 1922–7', *First World War Studies*, 2 (2011), 75–107.

136. 'Midlands' Two Minutes Silence', *Birmingham Gazette*, 12 November 1920, 3. This is a long and proud report about how widely and respectfully Midlands towns had carried out the act of remembrance.

137. Quoted in Geoff Dyer, *The Missing of the Somme* (Edinburgh: Canongate Books, 2016), 25–6.

138. 'Armistice Silence', *Edinburgh Evening News*, 10 December 1924, 6. Letter by James Clemenson, 'The Armistice Silence', *Westminster Gazette*, 14 November 1925, 4.

139. Winter, 'Thinking About Silence', in E. Ben-Ze'ev, *Shadows of War*, 3–31.

140. Cowgill, 'Canonizing Remembrance', 77.

141. Cowgill, 'Canonizing Remembrance', 78.

142. H. H. Thompson quoted in Gregory, *Silence of Memory*, 135.

143. Hynes, *A War Imagined*, ix–xi.

Chapter 3

1. From the chapter called 'The Power in the Air', in J. W. Robertson Scott, *England's Green and Pleasant Land* (Harmondsworth: Penguin, 1925), 94.

2. Asa Briggs, *The History of Broadcasting in the United Kingdom*, 1, *The Birth of Broadcasting* (Oxford University Press, 2000), 17.

3. Paddy Scannell and David Cardiff, *A Social History of British Broadcasting: 1922–1939, Serving the Nation* (Oxford: Basil Blackwell, 1991), 8; Paddy Scannell, *Radio, Television and Modern Life* (Oxford: Blackwell, 1996), 23–4; David Hendy, *Public Service Broadcasting* (Basingstoke: Palgrave Macmillan, 2013), 17.

4. David Hendy, 'The Great War and British Broadcasting. Emotional Life in the Creation of the BBC', *New Formations*, 82 (2014): 84.

5. Henry Thompson, 'Telephone London', in *Living London*, ed. George Sims (London: Cassell, 1901), 3: Sect. 1, 115.

6. John Reith, *Broadcast over Britain* (London: Hodder & Stoughton, 1924), 49.

7. Reith, *Broadcast over Britain*, 51.

8. Reith, *Broadcast over Britain*, 51, 54.

9. Iain Logie Baird, 'Capturing the Song of the Nightingale', *Science Museum Journal*, 4 (2015), accessed 6 September 2017, http://journal.sciencemuseum.ac.uk/browse/issue-04/capturing-the-song-of-the-nightingale/; Jeremy Mynott, *Birdscapes: Birds in Our Imagination and Experience* (Princeton, NJ: Princeton University Press, 2009), 178–80; Mark Rothenberg, *Why Birds Sing* (London: Penguin, 2006), 142–3; Richard Mabey, *Whistling in the Dark: In Pursuit of the Nightingale* (London: Sinclair-Stevenson, 1993), 99–111. Kate Lacey and David Hendy have touched on its significance to early broadcasting. Lacey, *Listening Publics*, 81–2; Hendy, 'The Great War and British Broadcasting', 92. The BBC's website includes a recording as part of its institutional history, 'Beatrice Harrison, Cello and Nightingale Duet 19 May 1924', accessed 23 May 2016, http://www.bbc.co.uk/programmes/p01z12h7.

10. Briggs, *The Birth of Broadcasting*, 127.

11. James Curran and Jean Seaton, *Power Without Responsibility: Press and Broadcasting in Britain* (London: Routledge, 1997), 111–12.

12. Hendy argues that Reith recognized his own psychological weaknesses, reading Freud as well as the scriptures: *Public Service Broadcasting*, 20–21. Ian McKintyre's biography does not attempt to explain Reith's personality in relation to his early background but does, by working with Reith's diaries, expose for example his very un-Reithian love affairs: *The Expense of Glory: A Life of John Reith* (London: HarperCollins, 1993).

13. John Reith, *Wearing Spurs* (London: Hutchinson, 1966), 179.

14. John Reith, *Into the Wind* (London: Hodder & Stoughton, 1949), 28.

15. Reith, *Into the Wind*, 103.

16. The 'broadcasting craze' was already frazzling the nerves of some in the very first year of the BBC's operations. It was 'that disturber of the peace in our homes, hotels, tea-shops, shaving saloons, and railway trains. The stress and excitement of modern life are enough to rattle the nerves of the strongest of us, and we need peace and quiet as restoratives': letter, *John O' London's Weekly*, 11 August 1923, 625.

17. Reith, *Broadcast over Britain*, 15–17.
18. Wilfred Whitten, 'The Lure and Fear of Broadcasting', *John O' London's Weekly*, 15 March 1924, 865.
19. Whitten, 'The Lure and Fear.'
20. Whitten, 'The Lure and Fear.'
21. John Reith, 'The Lure and Fear of Broadcasting. A Reply to John O' London', *John O' London's Weekly*, 22 March 1924, 938.
22. Arthur Burrows, 'Broadcasting the Nightingale', *Radio Times*, 14 December 1923, 428. 'The Londonderry Air' hailed from the Irish county of Derry and was used as the setting for the song 'Danny Boy' written in 1913. It was a wartime favourite and the tune made one of the most popular songs of the era by Elsie Griffin. 'Danny Boy' was a patriotic song of love, longing, and summer meadows. See Helen O'Shea, 'Defining the Nation and Confining the Musician: The Case of Irish Traditional Music', *Music and Politics*, 3/2 (2009), https://doi.org/10.3998/mp.9460447.0003.205.
23. Patricia Cleveland-Peck, *The Cello and the Nightingales: The Autobiography of Beatrice Harrison* (London: John Murray, 1985), 128–31.
24. The Savoy Orpheans were the house band at the Savoy Hotel, which was adjacent to the BBC premises on Savoy Hill. This meant that regular live broadcasting from the hotel was relatively straightforward in technical terms. Broadcasting from the Surrey woods was far more complex and involved setting up several microphones in the trees and a listening booth, and linking up to the telephone lines to be able to send the live sound to London for onward broadcast.
25. Cleveland-Peck, *The Cello and the Nightingales*, 133. There were just over a million licence holders at the end of 1924, but there could be several listening at once and an unknown group of unlicensed listeners too. See Briggs, *The Birth of Broadcasting*, 17.
26. Harrison Sisters' Collection, Royal College of Music, London, box 224. These are letters and cards from 1924 to 1927 and are presumably edited highlights.
27. Mark Cocker and Richard Mabey, *Birds Britannica* (London: Chatto & Windus, 2005), 340.
28. R. M. Monk, letter, Harrison Sisters' Collection, Royal College of Music, London, box 224.
29. W. J. Daully, letter, Harrison Sisters' Collection.
30. Asa Briggs, *The History of Broadcasting in the United Kingdom*, 2, *The Golden Age of Wireless* (London: Oxford University Press, 1965), 73.
31. John B. Thompson, *The Media and Modernity: A Social Theory of the Media* (Cambridge: Polity Press, 2001), 200.
32. Scannell and Cardiff, *A Social History of British Broadcasting*, 356–8. The enjoyment of music was diminished by the low fidelity of the apparatus and 'oscillation', which could produce sounds 'like the chirping of crickets': J. A. Fleming, 'The Polite Use of the Ether', *Radio Times*, 2 July 1926, 41.
33. John Reith, 'When Silence Was Served', *Radio Times*, 12 June 1925, 529–30.
34. 'Listening-in', *Bristol Times and Mirror*, 22 May 1924, 6.
35. 'A Word for the Nightingale', *Birmingham Daily Mail*, 24 May 1924.

36. Reith, 'When Silence Was Served', 530. For a discussion of a more recent controversy about the authenticity of the nightingale performances see Mynott, *Birdscapes*, 312–17.

37. John Blunt, 'Nightingales and Headphones', *Daily Mail*, 20 May 1924, 7; John Reith, 'Concerning Tinned Nightingale', *Radio Times*, 11 April 1924, 85–6.

38. Reith, *Broadcast over Britain*, 129. For a discussion of the standardization debates in the 1930s, see Briggs, *The Golden Age*, 39.

39. John Reith, 'The Broadcasting of Silence', *Radio Times*, 6 June 1924, 437.

40. Reith, 'The Broadcasting of Silence'.

41. Reith, *Broadcast over Britain*, 147.

42. Reith, *Broadcast over Britain*, 221.

43. Sara Maitland writes that birds are not silent, 'but still they somehow inhabit the spaces of silence. You need to be silent to see them, and they come and go as a silent gift': *A Book of Silence* (London: Granta, 2009), 160.

44. See Hendy's discussion of broadcasting's capability to unbalance minds and outlooks: 'The Great War and British Broadcasting', 101.

45. Editorial, 'Broadcasting Birds', *Times*, 21 May 1924, 15.

46. Robert Graves and Alan Hodge, *The Long Week-End: A Social History of Great Britain 1918–1939* (Harmondsworth: Penguin, 1971), 234.

47. Pulbrook, *The English Countryside*, 106. My emphasis.

48. See, for example, David James and Philip Tew, *New Versions of Pastoral: Post-Romantic, Modern, and Contemporary Responses to the Tradition* (Madison, NJ: Fairleigh Dickinson University Press, 2009); Harris, *Romantic Moderns*; Saylor, *English Pastoral Music*.

49. Reith, *Broadcast over Britain*, 217.

50. Matthew Arnold, *Culture and Anarchy and Other Writings*, ed. Stefan Collini (Cambridge: Cambridge University Press, 2002), 111, 122. On Arnoldian Reithianism, see Hendy, *Public Service Broadcasting*, 14–16.

51. Classless culture was conceived of as middle-class, in as much as everyone was entitled to the best. See Arnold, *Culture and Anarchy and Other Writings*, 79.

52. On mass culture and English society, see Geoffrey Searle, *A New England? Peace and War, 1886–1918* (Oxford: Oxford University Press, 2004), 107–15, 530–70; Hendy, *Public Service Broadcasting*, 13–15.

53. Wilfred Whitten, 'Pray Silence', *John O' London's Weekly*, 21 July 1923, 517.

54. Graves, *Goodbye to All That*, 236. Graves and Hodge reported that 'in most cases the blood was not running pure again for four or five years; and in numerous cases men who had managed to avoid a nervous breakdown during the war collapsed badly in 1921 or 1922': *The Long Week-End*, 23.

55. Hendy, *Public Service Broadcasting*, 12.

56. Hendy, 'The Great War and British Broadcasting', 84.

57. Overy, *The Morbid Age*; David Reynolds, *The Long Shadow: The Great War and the Twentieth Century* (London: Simon and Schuster, 2014), 141–5.

58. Reith, 'The Broadcasting of Silence', 347.

59. Reith, 'The Broadcasting of Silence', 347.

60. Mabey, *Whistling in the Dark*, 103.
61. Sam Bonner, audio recording, 1924, British Library catalogue number C653/3. Sam Bonner, Senior Duty Officer at the BBC's London Control Room in 1942, says in this recording that the 'bird and bomber-force were equal in volume' when he requested a recording line to be established and advised against transmission.
62. *The Demi-Paradise*, 1943, dir. Anthony Asquith.
63. Cleveland-Peck, *The Cello and the Nightingales*, 162.
64. David Elliston Allen, *The Naturalist in Britain: A Social History* (Princeton, NJ: Princeton University Press, 1994), 183, 210.
65. Moss, *A Bird in the Bush*, 122.
66. Briggs, *The Birth of Broadcasting*, 262.
67. Peter Eckersley, 'Broadcasting the Zoo', *Radio Times*, 22 August 1924, 374.
68. Arthur Burrows, *The Story of Broadcasting* (London: Cassell, 1924), 262.
69. Edward C. Ash, 'Broadcasting Wild Fowl at Night', *Radio Times*, 23 January 1925, 197–8.
70. Reith, *Broadcast over Britain*, 152.
71. E. Kay Robinson's beliefs are apparent in several of his books and pamphlets: *The Country Day by Day* (London: William Heinemann, 1905); *Religion of Nature* (London: Hodder and Stoughton, 1906); *The Meaning of Life* (London: Hampton Wick, 1916). On the wireless, Robinson gave eighteen talks in 1923 and 1924 rich in the sensory impressions of nature, which Reith liked enough to collect in the new Broadcast Library Series with a very personal introduction from himself. See E. Kay Robinson, *At Home with Nature* (London: Hodder and Stoughton, 1924).
72. Burrows, 'Broadcasting the Nightingale', 428.
73. 'When Will Insects Broadcast? The Ultra-Microphone and Its Wonders', *Radio Times*, 14 March 1924, 442.
74. Quoted in Baird, 'Capturing the Song of the Nightingale.'
75. W. H. Preece, 'The Phonograph', *Journal of the Society of the Arts*, 26 (1878).
76. 'The Microphone', *Spectator*, 25 May, 1878, 10.
77. Todd Avery argues that Reith felt a moral duty to promote his 'Christian ethics' as a necessary part of citizenship and national culture: *Radio Modernism: Literature, Ethics, and the BBC, 1922–1938* (Aldershot: Ashgate, 2006), 7–19.
78. Peter Bowler, *Science for All: The Popularisation of Science in Early Twentieth-Century Britain* (Chicago: University of Chicago Press, 2009), 34.
79. Bowler, *Science for All*, 34–5.
80. Bowler, *Science for All*, 41.
81. Leigh Eric Schmidt, *Hearing Things: Religion, Illusion, and the American Enlightenment* (Cambridge, MA: Harvard University Press, 2000), 238.
82. Bowler, *Science for All*, 233.
83. *Radio Times*, June to September 1924.
84. Bowler, *Science for All*, 44–8.
85. Burrows, *The Story of Broadcasting*, 178. Burrows borrows from George Eliot's *Middlemarch*: 'If we had a keen vision and feeling of all ordinary human life,' it

would be like hearing the grass grow and the squirrel's heart beat, and we should die of that roar which lies on the other side of silence. As it is, the quickest of us walk about well wadded with stupidity.'

86. Alfred Noyes, 'Radio and the Master-Secret', *Radio Times*, 18 September 1925, 550. D. L. LeMahieu argues that this article by Noyes is part of the BBC's campaign to emphasize the social and cultural benefits of the novel medium of communication, to gain legitimacy with the public, a phase that tailed off from 1927: *A Culture for Democracy: Mass Communication and the Cultivated Mind in Britain Between the Wars* (Oxford: Clarendon, 1988), 180–2.

87. Reith, *Broadcast over Britain*, 218–19.

88. Oliver Lodge, *Ether and Reality: A Series of Discourses on the Many Functions of the Ether of Space* (London: Hodder & Stoughton, 1925), 154.

89. Lodge, *Ether and Reality*, 236.

90. *Radio Times*, January to March 1925.

91. Reith, *Broadcast over Britain*, 223.

92. Anthony Enns, 'Psychic Radio: Sound Technologies, Ether Bodies, and Vibrations of the Soul', *The Senses and Society*, 3 (2008), 144–5.

93. Oliver Lodge, *Talks about Wireless: With Some Pioneering History and Some Hints and Calculations for Wireless Amateurs* (London: Cassell, 1925), 243.

94. Reith looked forward to 'world-unity' and 'peace on earth' at the end of his book, and in 1927 when he became director general of the Corporation he put forward the motto 'Nation shall speak peace unto nation' to underline the scope of broadcasting to promote global fellowship. See Reith, *Broadcast over Britain*, 222. John Durham Peters has argued that the vision of a universe held together by a transcendent, invisible principle of order (the ether) was an idea that resonated with a conservative social outlook: *Speaking into the Air: A History of the Idea of Communication* (Chicago: University of Chicago Press, 1999), 101.

95. See Penelope Gouk, 'The Harmonic Roots of Newtonian Science', in *Let Newton Be!*, ed. J. Fauvel, R. Flood, M. Shortland, and R. Williams (Oxford: Oxford University Press, 1988), 102–25.

96. Reith, *Broadcast over Britain*, 223–4.

97. Reith, 'Broadcasting Silence', *Radio Times*, 6 June 1924, 348.

98. Reith, *Broadcast over Britain*, 224.

99. Reith, *Broadcast over Britain*, 221.

100. Lodge published *Raymond or Life and Death* in 1916, an account of his communication with his son who had died in WWI. See W. P. Jolly, *Oliver Lodge* (Rutherford: Fairleigh Dickinson University Press, 1975), 205.

101. See Diarmaid MacCulloch, *Silence: A Christian History* (London: Allen Lane, 2013), 223–4; Maitland, *A Book of Silence*, 116–53.

102. Whitten, 'Pray Silence', 517–18.

103. Reith, *Broadcast over Britain*, 196–7.

104. Lacey has argued that for a public broadcaster devoted to the listener's self-improvement, silence was part of the scheduled flow of the week, not just a Sunday convention: *Listening Publics*, 82. The BBC included silent broadcasting breaks of

about fifteen minutes between programmes throughout the week, to allow listeners to switch off or digest what they had heard or get ready for what they were about to hear.

105. Reith, *Broadcast over Britain*, 196.
106. Briggs singles out the BBC's religious policy of the 1920s as 'standing out against the many of the tendencies of the age': *The Golden Age*, 7.
107. Burrows, *The Story of Broadcasting*, no pagination.
108. Burrows, *The Story of Broadcasting*, 75.
109. 'Come, Birdy, Come!', *Popular Wireless*, 3 June 1922, 7.
110. P. J. Risdon, 'Nature's Wireless', *Popular Wireless*, 22 March 1924, 123.
111. Burrows, *The Story of Broadcasting*, 122.
112. Letter, 'Does Listening Promote Health?', *Radio Times*, 23 November 1923, 320.
113. 'Wireless and Health. How Listening Affects Your Well-Being', *Radio Times*, 11 January 1924, 82.
114. 'Radio Rest Cures', *Popular Wireless*, 19 April 1924, 262.
115. Sir Bruce Bruce-Porter, 'Health and Headphones', *Radio Times*, 31 July 1925, 1.
116. Lord Knutsford, 'Wireless for the Wards', *Radio Times*, 5 June 1925, 504.
117. Lewis J. Ferrars, 'How We Wireless', *The Broadcaster*, June 1923, 46.
118. J. F. Corrigan, 'Aerials in Miniature', *Wireless World*, 1 May 1926, 385–6. Note that insect antennae were soon shown to do no such thing—they were found to be chiefly touch organs. For another example of nature's wireless, see E. de Poynton, 'A Mystery of the Animal World. Nature's Own Wireless', *Radio Times*, 12 December 1924, 526.
119. 'A Sea Shell Loud Speaker', *Popular Wireless*, 24 November 1923, 471.
120. Burrows, *The Story of Broadcasting*, 1, 5.
121. Burrows, *The Story of Broadcasting*, 3–5.
122. See Lacey, *Listening Publics*, 118–19.
123. Summer was a slack period for receiver sales. See Keith Geddes, *The Setmakers: A History of the Radio and Television Industry* (London: BREMA, 1991), 38.
124. Russell Mallinson, 'The Sunny Side of Radio', *The Broadcaster*, August 1923, 18.
125. Jonathan Hill suggests these machines could weigh up to 35 kg: *Radio! Radio!* (Bampton: Sunrise, 1986), 54. However, the basic crystal set reigned supreme over the significantly more expensive valve set for the first five years of BBC broadcasting: Geddes, *The Setmakers*, 16.
126. Though wireless could be more of a source of domestic strife than harmony, the BBC wanted the listening habit to restore a more rational use of leisure time and preserve the ideology of home, health and family, so important, it thought, to the nation: Scannell and Cardiff, *A Social History of British Broadcasting*, 358–69. See also Michael Bailey, 'The Angel in the Ether. Early Radio and the Constitution of the Household', in *Narrating Media History*, ed. Michael Bailey (London: Routledge, 2009), 52–65.
127. Quoted in Scannell and Cardiff, *A Social History of British Broadcasting*, 369.
128. Lacey, *Listening Publics*, 37.
129. Colls, *Identity of England*, 224.
130. W. M. Childs, *Holidays in Tents* (London: J. M. Dent, 1921), 27.

131. Sinclair Russell, 'The Radio Pipes of Pan', *The Broadcaster*, July 1923, 59. Many commentators saw this kind of movement as an invasion of the countryside by the vulgar, noisy masses. See, for example, John Carey, *The Intellectuals and the Masses: Pride and Prejudice Among the Literary Intelligentsia 1880–1939* (London: Faber and Faber, 1992), 46–92; Jon Agar, 'Bodies, Machines, Noise', in *Bodies/Machines*, ed. Iwan Rhys Morus (Oxford: Berg, 2002).

132. Reith, 'The Broadcasting of Silence', 437. For an example of the use of the word 'romantic', see a reflection in 1930 on the first years of broadcasting: Hervey Elwes, 'Five Years' Broadcasting', *Vox*, 11 January 1930, 334.

133. Gordon Selfridge, 'Mr Selfridge Expresses his Views', *Popular Wireless*, 3 June 1922, 7.

134. Russell, 'The Radio Pipes of Pan', 60.

135. Russell Mallinson, 'Radio and the Broad Highway', *The Broadcaster*, June 1923, 24.

136. Mallinson, 'The Sunny Side of Radio', 18. For an insight into the potential for acoustic drama of outdoor spaces in earlier times, see Bruce R. Smith, *The Acoustic World of Early Modern England: Attending to the O-Factor* (Chicago: Chicago University Press, 1999), especially 76–7.

137. Scannell and Cardiff, *A Social History of British Broadcasting*, 356.

138. Peters has discussed the possible origins of the term broadcasting: *Speaking into the Air*, 207. Whitten described Reith as a 'gentleman farmer' in his work as a broadcaster: 'A Word About Mr Broadcast', *John O' London's Weekly*, 13 December 1924, 433. Eric Gill's sculpture 'The Sower', in the main reception of Broadcasting House, was commissioned by Reith in 1930.

Chapter 4

1. Walter Benjamin, 'A Berlin Chronicle', in *Walter Benjamin: Selected Writings, Volume 2: Part 2, 1931–1934*, ed. Michael W. Jennings, Howard Eiland and Gary Smith (Cambridge, MA: Harvard University Press, 2005), 598.

2. For example, Nicholas J. Saunders and Paul Cornish, eds., *Modern Conflict and the Senses* (London, Routledge, 2017) and Mark M. Smith, *The Smell of Battle, the Taste of Siege: A Sensory History of the Civil War* (New York: Oxford University Press, 2015). A broader study of the senses is found in the six-volume set Constance Classen, ed., *The Cultural History of the Senses* (London: Bloomsbury, 2014).

3. On the sensuality of touch, including a consideration of touching natural things, see Yi-Fu Tuan, 'The Pleasures of Touch', in *The Book of Touch*, ed. Constance Classen (Abingdon: Routledge, 2020), 74–79.

4. Maurice Merleau-Ponty's concept of the body as 'the vehicle of being in the world' prompts the consideration that any one sense when called to attention brings the concordant operation of others towards the common goal of understanding the environment: *Phenomenology of Perception*, trans. C. Smith (London: Routledge, 1992), 81–2. See also Tim Ingold, 'Stop, Look, and Listen! Vision, Hearing, and Human

Movement', in *The Perception of the Environment: Essays on Livelihood, Dwelling, and Skill* (London: Routledge, 2000), 262.

5. Engagement with the natural world as a modern practice is explored in Jeremy Burchardt, 'Editorial: Rurality, Modernity and National Identity between the Wars', *Rural History*, 21 (2010), 143–50; Tim Edinsor, 'The Social Life of the Senses: Ordering and Disordering the Modern Sensorium', in *The Cultural History of the Senses in the Modern Age*, ed. David Howes (London: Bloomsbury, 2019), 31–54; Matless, *Landscape and Englishness*.

6. Tom Stephenson, *The Countryside Companion* (London: Odhams, 1939), 26.

7. Some of the best syntheses are Howard Hill, *Freedom to Roam: The Struggle for Access to Britain's Moors and Mountains* (Ashbourne, Derbyshire: Moorland, 1980); Marion Shoard, *A Right to Roam* (Oxford: Oxford University Press, 1999); Tom Stephenson, *Forbidden Land* (Manchester: Manchester University Press, 1989); and Harvey Taylor, *A Claim on the Countryside: A History of the British Outdoor Movement* (Edinburgh: Keele University Press, 1997).

8. Matless, *Landscape and Englishness*, 62–100; Agar, 'Bodies, Machines and Noise', 197–220; Paul Brassley, Jeremy Burchardt, and Lynne Thompson, *The English Countryside Between the Wars: Regeneration Or Decline?* (Woodbridge: Boydell Press, 2006); John Sheail, *Rural Conservation in Interwar Britain* (Oxford: Oxford University Press, 1981).

9. The terms appear in a key preservationist texts of the time such as Clough Williams-Ellis, *Britain and the Beast* (London: J. M. Dent, 1937) and in C. E. M. Joad, *The Untutored Townsman's Invasion of the Countryside* (London: Faber and Faber, 1946). On the Romantic gaze, see Jon Urry, 'The Tourist Gaze "Revisited"', *American Behavioural Scientist*, 36 (1992), 172–86.

10. Rebecca Solnit, *Wanderlust: A History of Walking* (London: Granta, 2014), 120, 164.

11. On the social mix, see Ann Holt, 'Hikers and Ramblers: Surviving a Thirties' Fashion', *The International Journal of the History of Sport*, 4 (1987), 59–60. On women, see Claire Langhamer, *Women's Leisure in England, 1920–60* (Manchester: Manchester University Press, 2000), 76–9.

12. Graves and Hodge, *The Long Week-End*, 271. See also Claude Fisher, 'And Now for the Open Road!', *Daily Mail*, 21 March 1932, 10. The 'craze' is reported to be declining by the end of 1936: 'Rambling as a Fashion', *Manchester Guardian*, 1 December 1936, 13.

13. 'Escape from Squalor: Rambling the Best Way', *Manchester Guardian*, 19 June 1933, 13.

14. Taylor, *Claim on the Countryside*, 234.

15. Hill, *Freedom to Roam*, 13–17; David Prynn, 'The Clarion Clubs, Rambling and the Holiday Associations in Britain since the 1890s', *Journal of Contemporary History*, 11 (1976), 65.

16. 'Rambling an Art: Stemming the Tide of Mechanicalism', *Manchester Guardian*, 2 October 1933, 2.

17. On the problem of urban noise and other modern city pressures see Snaith, *Sound and Literature*; Mansell, *The Age of Noise in Britain*; Stephen Kern, *The Culture of*

Time and Space 1880–1918 (Cambridge, MA: Harvard University Press, 2003); and Emily Cockayne, *Hubbub, Filth, Noise & Stench in England 1600–1770* (New Haven, CT: Yale University Press, 2008).

18. Tim Edensor develops similar arguments in 'Walking in the British Countryside: Reflexivity, Embodied Practices and Ways to Escape', *Body & Society*, 6 (2000), 81–106.

19. Mark Jackson, *The Age of Stress: Science and the Search for Stability* (Oxford: Oxford University Press, 2013).

20. Catherine Gayler, diary (1934), Bishopsgate Institute, London, catalogue number GDP/16.

21. G. H. B. Ward, Sheffield Clarion Ramblers handbook, 1921–22, quoted in Dave Sissons, Terry Howard, and Roly Smith, *Clarion Call: Sheffield's Access Pioneers* (Sheffield: Clarion Call, 2017), 14.

22. This is a term that appears in the Manchester Ramblers' Federation handbook, quoted in Ben Anderson, 'A liberal countryside? The Manchester Ramblers' Federation and the "social readjustment" of urban citizens, 1929–1936', *Urban History*, 38 (2011), 86.

23. On Ewan MacColl's song writing see Ben Harker, ' "The Manchester Rambler": Ewan MacColl and the 1932 Mass Trespass', *History Workshop Journal*, 59 (2005), 219–28.

24. Hill, *Freedom to Roam*, 53.

25. Taylor, *Claim on the Countryside*, 253.

26. Tom Hall, *Citizen Rambles* (Glasgow: James Hedderwick, 1929), 7.

27. Peters, *Speaking Into the Air*, 160.

28. George Orwell, *The Road to Wigan Pier*, in *Orwell's England*, ed. Peter Davison (London: Penguin, 2001), 197.

29. Arnold Haultain, *Of Walking and Walking Tours. An Attempt to Find a Philosophy and Creed* (London: T. Werner Laurie, 1914), 199.

30. Haultain, *Of Walking and Walking Tours*, 31.

31. R. J. Moore-Colyer, 'Great Wen to Toad Hall: Aspects of the Urban-Rural Divide in Inter-War Britain', *Rural History*, 10 (1999), 107. See also Howkins, *The Death of Rural England*.

32. Tristram Hunt, *Making Our Mark* (London: Campaign to Protect Rural England, 2006), 6.

33. The classic rallying cry for action against suburban sprawl was written by the architect, urban planner and preservationist Clough Williams-Ellis: *England and the Octopus* (London: Geoffrey Bles, 1929).

34. Williams-Ellis, *England and the Octopus*, 128: 'The way in which many of our larger towns have allowed their main approaches to be devastated by advertising is truly surprising. Some of them devote large sums to the maintenance of civil dignity at their centres, but leave their boundaries to be as tatterdemalion and slattern as they please.'

35. Richard Hornsey, ' "He Who Thinks, in Modern Traffic, is Lost": Automation and the Pedestrian Rhythms of Interwar London', in *Geographies of Rhythm: Nature, Place, Mobilities and Bodies*, ed. Tim Edensor (Farnham: Ashgate, 2010), 105.

36. Hornsey, ' "He Who Thinks" ', 99, 101.

37. Graves and Hodges, *The Long Week-End*, 291–2.

38. The key history of dance culture at this time is James Nott, *Going to the Palais: A Social And Cultural History of Dancing and Dance Halls in Britain, 1918–1960* (Oxford: Oxford University Press, 2015).

39. Leslie Stephen's essay 'In Praise of Walking' is collected in *Studies of a Biographer* (London: Duckworth, 1902), 256.

40. Sydney Moorhouse, *Walking Tours and Hostels in England* (London: Country Life, 1936), xi.

41. G. M. Trevelyan's essay 'Walking' is collected in *Clio, A Muse and Other Essays Literary and Pedestrian* (London: Longmans, Green, 1913), 61.

42. Edgar Brooks, *Country Rambles Round Birmingham and Week-end Holidays for the Pedestrian, Cyclist, and Motorist*, 6th ed. (Birmingham: Cornish Brothers, 1927), 10–12.

43. Trevelyan, 'Walking', 69–71.

44. Trevelyan, 'Walking', 74.

45. Raphael Samuel tells the tales of his mother's walking in *Theatres of Memory, Volume 2: Island Stories. Unravelling Britain* (London: Verso, 1998), 132. Men also wanted to get out of the house—see Stephen Graham, *The Gentle Art of Tramping* (London: Bloomsbury 2019 [1927]), 26, who wrote that 'tramping is first of all a rebellion against housekeeping and daily and monthly accounts'.

46. Geoffrey Murray, *The Gentle Art of Walking* (London: Blackie, 1939), 294.

47. Quoted in E. D. Laborde, *Popular Map Reading* (Cambridge: Cambridge University Press, 1928), 94.

48. Samuel, *Island Stories*, 133.

49. Trevelyan, 'Walking', 81.

50. 'Hiking Heroism in Yorkshire', *Leeds Mercury*, 24 November 1931, 6.

51. Claude Fisher, 'All the Year is Hiking Time', *Daily Mail*, 15 September 1932, 4.

52. Stan Oxley and Willis Marshall's journal, 'Our First Walking Tour, 4 to 11 Aug 1923', Chapter VI.

53. The journals are kept by Marshall's family today. There are two: Willis Marshall, 'Rambles from Sheffield into Derbyshire, January 21st 1923 to September 28th 1924' and Stan Oxley and Willis Marshall, 'Our First Walking Tour, 4 to 11 Aug 1923'. Ann Beedham has reproduced some of the writing, photographs, and illustrations in *Days of Sunshine and Rain: Rambling in the 1920s* (Sheffield: You Books, 2011).

54. Oxley and Marshall 'Our First Walking Tour', Chapter 2.

55. This is the first half of the verse. Marshall, 'Rambles from Sheffield', Ramble 88, 2–9 August 1924.

56. Beedham, *Days of Sunshine and Rain*, 62.

57. Oxley and Marshall, 'Our First Walking Tour', Chapter 5.

58. Marshall, 'Rambles from Sheffield', Ramble 81, 14 to 15 June 1924.

59. 'Hikers Miss the Sun. Vain All Night Ramble to Sussex Beacon', *Portsmouth Evening News*, 12 September, 1932, 3. Mais led another unsuccessful trip to Chanctonbury Ring that year highlighted by Graves and Hodge, *The Long Week-End*, 271–2.

60. Claude Fisher, 'Take Your Holidays by Moonlight', *Daily Mail*, 18 August, 1933, 6.

61. Another example is C. R. W. Nevinson's poster for a bus company in 1921 to promote weekend walking that showed a well-dressed couple on a moonlit night emerging from or entering the woods.

62. Oxley and Marshall, 'Our First Walking Tour', Chapter 2.

63. Melanie Tebbutt, 'Rambling and Manly Identity in Derbyshire's Dark Peak, 1880s–1920s', *The Historical Journal*, 49 (2006), 1144–7, 1150–1.

64. Oxley and Marshall, 'Our First Walking Tour', Chapter IV.

65. Trevelyan, 'Walking', 70.

66. Jack Jordan writing in the Sheffield Clarion Ramblers handbook, 1958–9, quoted in Sissons, *Clarion Call*, 30.

67. Tebbutt, 'Rambling and Manly Identity', 1141.

68. Sissons, *Clarion Call*, 30. There was a Whitmanesque 'combination of fresh air, exercise, sunshine, nudity and male comradeship' in Ward's writing and thought according to Sissons, quoted in Tebbutt, 'Rambling and Manly Identity', 1146. Carpenter was a philosopher and charismatic propagandist for a multitude of causes, including liberalizing attitudes to homosexuality along with women's liberation, naturism, vegetarianism, animal rights, smallholdings, and environmentalism.

69. C. E. M. Joad, *A Charter for Ramblers: The Future of the Countryside* (London: Hutchinson, 1934), 150. Joad recommended nude sea bathing on deserted beaches and drying 'naked to the sun' as Matless points out in his discussion of the contradictions and politics associated with English and German nudism: *Landscape and Englishness*, 95–8.

70. Orwell, *The Road to Wigan Pier*, 175: 'One sometimes gets the impression that the mere words "Socialism" and "Communism" draw towards them with magnetic force every fruit-juice drinker, nudist, sandal-wearer, sex-maniac, Quaker, "Nature Cure" quack, pacifist and feminist in England.'

71. Arthur B. Craven, 'The Yorkshire Ramblers' Club, 1892–1992', *Yorkshire Ramblers' Club Journal*, 11 (1992), 1–5.

72. Oxley and Marshall 'Our First Walking Tour', Chapter 2.

73. On Ward's ideas about the masculinity of the landscape and of the rambler, see Tebbutt, 'Rambling and Manly Identity', especially at 1114; Melanie Tebbutt, 'Landscapes of Loss: Moorlands, Manliness and the First World War', *Landscapes* 5 (2004), especially at 116 and 125.

74. From a letter written by Oxley to Marshall in 1973, quoted in Beedham, *Days of Sunshine and Rain*, 23.

75. Oxley and Marshall 'Our First Walking Tour', Chapter 5.

76. Quoted in Sissons, *Clarion Call*, 29.

77. Charlotte Peacock, *Into the Mountain: A Life of Nan Shepherd* (Cambridge: Galileo, 2018), 210–17.

78. Nan Shepherd, *The Living Mountain* (London: Canongate, 2014 [1977]), 1.

79. *The Living Mountain* was not published until 1977. It is a difficult book to describe. Robert Macfarlane (*The Living Mountain*, xii) suggests: 'A celebratory prose-poem?

A geo-poetic quest? A place-paean? A philosophical enquiry into the nature of knowledge? A metaphysical mash-up of Presbyterianism and the Tao?'

80. Peacock, *Into the Mountain*, 269.
81. Shepherd, *The Living Mountain*, 105.
82. Shepherd, *The Living Mountain*, 105.
83. Shepherd, *The Living Mountain*, 106.
84. Shepherd, *The Living Mountain*, 106.
85. Shepherd, *The Living Mountain*, 16.
86. Shepherd, *The Living Mountain*, 12–13.
87. Shepherd, *The Living Mountain*, 22.
88. Shepherd, *The Living Mountain*, 26.
89. Shepherd, *The Living Mountain*, 107.
90. Shepherd, *The Living Mountain*, 107.
91. Shepherd, *The Living Mountain*, 102.
92. Shepherd, *The Living Mountain*, 52.
93. Shepherd, *The Living Mountain*, 51.
94. Shepherd, *The Living Mountain*, 61.
95. Shepherd, *The Living Mountain*, 26.
96. Ward in the 1909 Sheffield Clarion Ramblers handbook, quoted in Tebbutt, 'Rambling and Manly Identity', 1141.
97. Dorothy Wordsworth's writing may well have been an influence for Shepherd, and Dorothy Pilley and Vita Sackville-West were contemporaries who wrote about their experiences of mountains and the natural world.
98. Shepherd, *The Living Mountain*, xxvii–xxix.
99. Although the European cultural hierarchy of senses assigned to women the low-ranking touch, taste, and smell, and to men the more elevated sight and hearing, I have found no clear gender distinctions relating to interests in or development of sensory experience. The classical distinctions have been explored by Constance Classen in 'Engendering Perception: Gender Ideologies and Sensory Hierarchies in Western History', *Body & Society*, 3 (1997), 1–19.
100. Wordsworth's influential place in walking culture is reviewed in Solnit, *Wanderlust*, 82, 94, 112.
101. Frederick Alexander Wills, *The Rambles of 'Vagabond' of the 'Newcastle Evening Chronicle'* (Newcastle Upon Tyne: J & P Bealls, 1936), 13–16. Mystical interpretations of the experience of nature could encompass the themes of universalism and continuity found in various spiritual beliefs or Christian theology. For an exploration of the interplay between nature mysticism and interwar modernity, see Matless, *Landscape and Englishness*, 84–86; David Matless, 'Nature, the Modern and the Mystic: Tales from Early Twentieth Century Geography', *Transactions of the Institute of British Geographers*, 16 (1991), 272–86.
102. Shepherd, *The Living Mountain*, 76.
103. Shepherd, *The Living Mountain*, 84–5.
104. Joad, *A Charter for Ramblers*, 55.
105. Shepherd, *The Living Mountain*, 107.

106. Frédéric Gros, *A Philosophy of Walking* (London: Verso, 2014), 7.

107. Letter, 'Ramblers' Conduct', *Sheffield Daily Telegraph*, 25 May 1928, 3.

108. Elyn Walshe, 'Green Roads: Solitude and the Litter Fiends', *Manchester Guardian*, 1 May 1933, 6.

109. Quoted in Holt, 'Hikers and Ramblers', 64. The poem is a parody of a popular Victorian reworking of Omar Khayyam's Rubáiyát by Edward FitzGerald.

110. C. E. M. Joad, *The Horrors of the Countryside* (London: Hogarth Press, 1931), 24–7; Aldous Huxley, 'The Outlook for American Culture: Some Reflections in a Machine Age', *Harper's Magazine*, August 1927, 265–70. Peter Bailey traces the heritage of this kind of narrative in *Leisure and Class in Victorian England: Rational Recreation and the Contest for Control, 1830–1885* (London: Routledge, 2007 [1978]). Also important is John Carey, *The Intellectuals and the Masses: Pride and Prejudice Among the Literary Intelligentsia 1880–1939* (London: Faber and Faber, 1992).

111. Letter, 'Anti-Desecration Crusade', *Sheffield Daily Telegraph*, 25 May 1928, 3.

112. For a brief but rich account of working-class charabanc excursion cultures of the 1930s and 40s, see Richard Hoggart, *The Uses of Literacy: Aspects of Working-Class Life* (London: Penguin, 2009 [1957]), 126–8.

113. From a 1938 parliamentary debate, quoted in Taylor, *Claim on the Countryside*, 230. According to Prynn, 'The Clarion Clubs', 70: 'In the 1930s tens of thousands of walkers seeking relief from idleness or dull work and dreary surroundings would pour into the Peak district from the neighbouring industrial cities.'

114. Peter Merriman, ' "Respect the Life of the Countryside": The Country Code, Government and the Conduct of Visitors to the Countryside in Post-War England and Wales', *Transactions of the Institute of British Geographers*, 30 (2005), 336–50.

115. Robert Louis Stevenson, *The Works of Robert Louis Stevenson*, Miscellanies, Volume III (London: Chatto & Windus, 1895), 174.

116. Graham, *Gentle Art of Tramping*, 28.

117. Graham, *Gentle Art of Tramping*, 91.

118. Arthur Sidgwick had interesting things to say about the rhythmic kinship between walking and music. He served with the Royal Garrison Artillery and was killed in action near Ypres on 17 September 1917. His book of walking essays was published before the war; A. H. Sidgwick, *Walking Essays* (London: Edward Arnold, 1912), 66–72.

119. Graham, *Gentle Art of Tramping*, 93.

120. A cheeky new version of this song became popular in the 1914–18 conflict, which carried the same tune to new lyrics: 'Kaiser Bill is feeling ill, / The Crown Prince, he's gone barmy. / We don't give a fuck for old von Kluck, / And all his bleeding army.' See Martin Pegler, *Soldiers' Songs and Slang of the Great War* (London: Osprey, 2014), 330.

121. Roy Palmer, *The Sound of History: Songs and Social Comment* (Oxford: Oxford University Press, 1988), 281.

122. Joad, *The Horrors of the Countryside*, 26.

123. Hoggart, *The Uses of Literacy*, 135–8.

124. A. L. Lloyd, *Folk Song in England* (St. Albans: Paladin, 1975), 298, 349. Lloyd refers to 'industrial folk songs' that were prevalent in northern towns and cities. On iron and steel workers' songs, see Palmer, *The Sound of History*, 89–92.
125. Clare V. J. Griffiths, *Labour and the Countryside: The Politics of Rural Britain 1918–1939* (Oxford: Oxford University Press, 2007), 95.
126. For example: G. H. B. Ward, *Songs for Ramblers to Sing on the Moorlands* (Sheffield: Sheffield Clarion Ramblers, 1922); B. H. Humble and W. M. McLellan, *Songs for Climbers* (Glasgow: W. M. McLellan, 1938); *The Hackney Scout Song Book* (London: Stacy & Son, 1921); and *Holiday Songs* (London: The Holiday Fellowship, 1937).
127. Graham, *The Gentle Art of Tramping*, 29.
128. See Ina Zweiniger-Bargielowska, *Managing the Body: Beauty, Health, and Fitness in Britain, 1880–1939* (Oxford: Oxford University Press, 2010), 309–315 and Helen Walker, 'The Popularisation of the Outdoor Movement, 1900–1940', *The British Journal of Sports History*, 2 (1985), 141, 151.

Chapter 5

1. Huxley, *Bird-Watching*, 116.
2. Overy, *The Morbid Age*, 316, 319. Overy argues that this mood existed without an obvious enemy in mind, though the Spanish Civil War was a catalyst in shaping British attitudes towards contemporary crisis.
3. On the importance of music during this war, see Christina Baade, *Victory Through Harmony: The BBC and Popular Music in World War II* (New York: Oxford University Press, 2012).
4. Scannell and Cardiff, *A Social History of British Broadcasting*, 277–303; Hajkowski, *The BBC and National Identity in Britain*.
5. Nicholas, *The Echo of War*, 2.
6. Koch's work has been studied chiefly for its natural history significance, with only Seán Street paying significant attention to his recordings as part of radio's culture: Street, *The Poetry of Radio*, 100–7; Street, *The Memory of Sound: Preserving the Sonic Past* (New York, Routledge, 2015), 123–4. Street argues that by the end of the 1940s Koch was a household name and as well-known to the public as David Attenborough is today: Ludwig Koch and the Music of Nature, BBC Radio 4, 15 April 2009, accessed 7 September 2019, https://www.bbc.co.uk/archive/ludwig-koch-and-the-music-of-nature/zd7kf4j. John Burton, who was in charge of the BBC wildlife sound library at the Natural History Unit from 1962 to 1988, has written about Koch's work as a foundation of the library's collection: 'The BBC Natural History Unit Wildlife Sound Library 1948-1988', *Wildlife Sound*, 12 (2012), 19–21. Others have looked briefly at how Koch's recording efforts gave scientists the opportunity to study birdsong in a new way, breaking away from problematic transcription traditions of old: Rothenberg,

Why Birds Sing, 58–60 and Joeri Bruyninckx, *Listening in the Field: Recording and the Science of Birdsong* (Cambridge, MA: MIT Press, 2018).

7. Briggs, *The Golden Age*, 99–100. Flat-disc recording was more flexible and became the prime method of recording and editing in the mid-1930s.

8. 'A Nest of Singing Birds', *Radio Times*, 13 November 1936, 72. Hawkins was a contributor to *Week-End Review*, *Twentieth Century* and *The Listener* and finding his feet as a novelist at this point, but went on immediately after the war to establish the BBC's Natural History Unit.

9. *His Master's Voice General Catalogue* (London: Gramophone Company, 1911), 67. The British wildlife photographer Cherry Kearton attempted to make recordings of a nightingale singing in the wild in 1900: John Bevis, *Direct from Nature: The Photographic Work of Richard & Cherry Kearton* (Axminster: Colin Sackett, 2007), 57.

10. Reich's recording can be compared to the recording of the Surrey nightingale singing free in 1927 by listening to the CD *Nightingales: A Celebration* (British Trust for Ornithology, 1998).

11. *His Master's Voice General Catalogue* (London: Gramophone Company, 1914), 139.

12. Peter Copeland, Jeffrey Boswall, and Leonard Petts, *Birdsongs on Old Records: A Coarsegroove Discography of Palearctic Region Bird Sound 1910-1958* (London: British Library Wildlife Sounds, National Sound Archive, 1988), 5.

13. Koch, *Memoirs of a Birdman*, 34.

14. Letter, 17 March 1936, BBC Written Archive, S26/1/1, June 1935–May 1936 correspondence.

15. Note, BBC Written Archive, Rconti1, Talks Tom Harrisson file1A 1932–1936.

16. Koch, *Memoirs of a Birdman*, 35.

17. Koch, *Memoirs of a Birdman*, 36.

18. Koch, *Memoirs of a Birdman*, 25.

19. Koch, *Memoirs of a Birdman*, 25–7.

20. Koch, *Memoirs of a Birdman*, 27–9.

21. Koch, *Memoirs of a Birdman*, 25.

22. Koch, *Memoirs of a Birdman*, 25. More than a decade earlier, Walter Garstang had published his affectionate little book *Songs of the Birds* (London: John Lane The Bodley Head, 1922) using these modes of representation, alongside his poems about the songs of some twenty birds. Before that, many poets had tried to capture the rhythms and sound shapes of the nightingale's voice, John Clare in 1832 rendering in words the sounds of the bird outside his window more accurately that anyone would for almost a century. See Rothenberg, *Why Birds Sing*, 24–5. For a detailed expansion about transcription, musical notation, and other methods of fixing birdsong see Mynott, *Birdscapes*, 160–70.

23. Koch, *Memoirs of a Birdman*, 26.

24. Koch, *Memoirs of a Birdman*, 17–20. Koch toured in Europe and was recorded as a musician and singer.

25. The BBC had not offered Koch the use of its own recording truck for 'outside broadcasts', which tended to be seen as a toy and rarely put into action. In 1936,

live transmissions of outside broadcasts made up only 2 per cent of national programmes: Briggs, *The Golden Age*, 54.

26. Nicholson and Koch, *More Songs of Wild Birds*, 36, 38. Nicholson commented on 'how the natural peace of the country was drowned under the indefatigable hum of distant engines and wheels'.

27. Nicholson and Koch, *Songs of Wild Birds*, xxvii.

28. Nicholson and Koch, *Songs of Wild Birds*, xxi.

29. Lacey has discussed how the privatization of listening to recorded sound and the radio did not necessarily create a disconnection from public culture: *Listening Publics*, 121. See also William Howland Kenney, *Recorded Music in American Life: The Phonograph and Popular Memory, 1890–1945* (New York: Oxford University Press, 2003). On the emergence of popular British birdwatching, see Moss, *A Bird in the Bush*, 86–149 and Michael Guida, '1928. Popular Bird-Watching Becomes Scientific: The First National Bird Census in Britain', *Public Understanding of Science*, 8 (2019), 622–7.

30. Julian Huxley, *Bird-Watching and Bird Behaviour* (London: Chatto & Windus, 1930), 7.

31. As early as 1916, Julian Huxley suggested that the 'vast army of birdlovers and bird-watchers to-day in existence' could be directed to channel their enthusiasms into solving, with biologists, fundamental problems of science: 'Bird-Watching and Biological Science (part 1)', *Auk*, 33 (1916), 142–61. For a discussion of Max Nicholson's shaping of the birdwatcher as citizen in the 1930s, see Macdonald, 'What Makes You a Scientist', 54–60.

32. Nicholson and Koch, *Songs of Wild Birds*, xxi.

33. On the idea of the objectivity of sound recording, see Lacey, *Listening Publics*, 57. See also Lorraine Daston and Peter Galison, *Objectivity* (Brooklyn, NY: Zone Books, 2007).

34. Nicholson and Koch, *Songs of Wild Birds*, xxii.

35. Nicholson and Koch, *More Songs of Wild Birds*, inside front cover.

36. Nicholson and Koch, *Songs of Wild Birds*, 197.

37. Nicholson and Koch, *Songs of Wild Birds*, 185.

38. Nicholson and Koch, *Songs of Wild Birds*, 186.

39. Nicholson and Koch, *Songs of Wild Birds*, xiii.

40. Nicholson and Koch, *Songs of Wild Birds*, xiv.

41. On acoustic transparency, see Sterne, *The Audible Past*, 256–9 and Lacey, *Listening Publics*, 67–71. The mass transit of minds from the domestic realm to somewhere better was a quality associated with radio at this time, notably through the production of mental imagery stimulated by a sense of proximity to a sound scene, argues Susan Douglas, *Listening In*, 26–30.

42. I can add my experience of listening to these recordings on a 1920s Columbia portable player. Listening to the recordings elicits surprise at the clarity and projection from a small machine of a tonal range that is concordant to my ears with the voices of some common birds such as the blackbird. This simple technology produces a representation of significant charm as well as convincing accuracy.

43. Witherby sales statement, December, 1936, BBC Written Archive, S26/1/6 Koch, correspondence file 6.

44. Book review, *British Birds*, 1 December 1937, 239.

45. Book Chronicle, *The Listener*, 4 November 1936, 877.

46. Quoted in Reynolds, *The Long Shadow*, 225. Military planners also held this view and they predicted 20,000 causalities in the initial twenty-four hours of an air attack, rising to perhaps 150,000 by the end of the first week. For an account of air raids and British culture that includes both world wars, see Susan Grayzel, *At Home and Under Fire: Air Raids and Culture in Britain from the Great War to the Blitz* (New York: Cambridge University Press, 2012). Historians have pointed out that mass air raid neurosis did not materialize and the durability of civilian morale has been shown by Edgar Jones, Robin Woolven, Simon Wessely, and Bill Durodie, 'Civilian Morale During the Second World War: Responses to Air Raids Re-Examined', *Social History of Medicine*, 17 (2004), 464–5.

47. Quoted in Reynolds, *The Long Shadow*, 226. Mansell (*The Age of Noise*, 161, 170) argues that prominent neurologist Sir James Purves-Stewart was among those who warned that the terrifying noises of the Blitz could bring about in urban populations an epidemic of shell shock of the kind seen in the trenches.

48. Brett Holman, *The Next War in the Air: Britain's Fear of the Bomber, 1908–1941* (Farnham and Burlington: Ashgate, 2014); Uri Bialer, *The Shadow of the Bomber: The Fear of Air Attack and British Politics, 1932–1939* (London: Royal Historical Society, 1980).

49. H. G. Wells's novel *The War in the Air* (1908) set the pattern for much of the fiction and non-fiction about the effects of aerial bombing.

50. Reynolds, *The Long Shadow*, 203.

51. Paul Nash, *Aerial Flowers* (Oxford: Counterpoint, 1947), 5.

52. Macdonald, 'What Makes You a Scientist', 56.

53. Seven million copies were sold according to Mynott, *Birdscapes*, 77.

54. R. A. Saville-Sneath, *Aircraft Recognition* (London: Penguin, 1990 [1941]), 32. Another 1941 book *The Spotter's Handbook* sold seven million copies (Macdonald, 'What Makes You a Scientist', 66).

55. H. Ramsden Whitty, *Observers' Tale: The Story of Group 17 of the R.O.C.* (London: Roland Brothers, 1951), 26.

56. *London Can Take It!*, 1940, dir. Harry Watt and Humphrey Jennings.

57. H. V. Morton, *H. V. Morton's London* (London: Methuen, 1948 [1942]), viii.

58. Morton, *H. V. Morton's London*, xv.

59. Photograph of demolition workers standing still and quiet while rescue workers listen, IWM, 'Air Raid Damage', catalogue number HU 680, accessed 29 May 2020, http://www.iwm.org.uk/collections/item/object/205068708.

60. Virginia Woolf, 'Thoughts on Peace in an Air Raid', *New Republic*, 21 October 1940, 549–51.

61. Tom Harrisson and Charles Madge, *War Begins at Home* (London: Chatto & Windus, 1940), 184–6 and see Joanna Bourke, *Fear: A Cultural History* (London: Virago, 2005), 229.

62. Harrisson and Madge, *War Begins at Home*, 201. This Mass Observation report extensively documents the largely negative feelings about the blackout for those at home and making their way about outdoors: see 186–203.

63. Quoted in Jo Fox, 'Careless Talk: Tensions Within British Domestic Propaganda During the Second World War', *The Journal of British Studies*, 51 (2012), 964. Fox has argued that the public reception to security propaganda that sought to curtail everyday talk was often one of irritation and resentment, because the natural urge to chat, a treasured national pastime and itself a part of maintaining morale, had been cast with suspicion.

64. Asa Briggs, *The History of Broadcasting in the United Kingdom*, vol. 3, *The War of Words* (Oxford: Oxford University Press, 1995), 95. News became the most listened-to programme, the nine o'clock news reaching an audience of as much as half of the population: see 48.

65. 'Influence of Air Raids on Listening Habits', October 1940, BBC Written Archive, R9/9/4 Audience Research Special Reports.

66. Scannell and Cardiff, *A Social History of British Broadcasting*, 221; Nicholas, *The Echo of War*, 51–3; David Cardiff and Paddy Scannell, '"Good Luck War Workers!" Class, Politics and Entertainment in Wartime Broadcasting', in *Popular Culture and Social Relations*, ed. Tony Bennett, Colin Mercer, and Janet Woollacott (Milton Keynes: Open University Press, 1986), 98.

67. Siân Nicholas, 'The People's Radio: The BBC and Its Audience, 1939–1945', in *'Millions Like Us?' British Culture in the Second World War*, ed. N. Hayes and J. Hill (Liverpool: Liverpool University Press, 1999), 77.

68. Nicholas, *The Echo of War*, 230. Vaughan Williams's music could be seen to be too English to represent the nation as a whole.

69. Nicholas, *The Echo of War*, 2. Many broadcasting scholars have argued that the BBC was a centre point for the construction and dissemination of British national identity during the 1930s and 40s. See Briggs, *The War of Words*, 95; Scannell and Cardiff, *A Social History of British Broadcasting*, 10; Hajkowski, *The BBC and National Identity in Britain*, especially 234.

70. The title for this section is taken from a poem about the BBC in *Punch*, quoted in Briggs, *The Golden Age*, 13.

71. Koch, *Memoirs of a Birdman*, 69.

72. Koch, *Memoirs of a Birdman*, 69.

73. See Richard Overy, *The Bombing War: Europe 1939–1945* (London: Penguin, 2014), 73–4.

74. Overy, *The Bombing War*, 192. For a description of the doodlebug attacks and their sound see the 1945 account by the Air Ministry's writer-in-residence, H. E. Bates, *Flying Bombs over England* (Oxford: ISIS, 1997, large print edition), 38–50. He describes the harshness of the sound followed by the terrible silence of its descent.

75. Bates, *Flying Bombs over England*, 45.

76. Ceri Richards was a musician, and he is almost certainly responding to and representing overhead sonic disturbances in this painting.

77. Overy, *The Bombing War*, 126–7.

78. Memorandum from John Polworth, 8 March 1946, BBC Written Archive, N2/25 North Region, *Country Magazine* file.
79. Koch, *Memoirs of a Birdman*, 71.
80. Nicholas, *The Echo of War*, 45.
81. 'Listen to Our Songsters', 13 June 1943, BBC Written Archive, Ludwig Koch scripts.
82. *Children's Hour* programme listing, *Radio Times*, 19 December 1941, 14.
83. 'The Nuthatch Sings in February', 7 February 1943, BBC Written Archive, Ludwig Koch scripts.
84. The engineer who stopped the broadcast retold the incident: Sam Bonner, audio recording, 1942, British Library catalogue number C653/3.
85. Koch's reputation after the war was ascending in such a way that in 1946 Desmond Hawkins, who went on to found the BBC Natural History Unit, created *The Naturalist* programme partly as a vehicle for Koch's recordings and thoughts about the natural world: Burton, 'The BBC Natural History Unit', 21.
86. Tom Harrisson, 'Radio' column, *Observer*, 3 January 1943, 2. Asa Briggs has said that Harrisson's *Observer* column offered 'a valuable weekly critique' of radio programmes, drawing upon his team of Mass Observation correspondents for opinion and his skills as a writer-producer himself: *The War of Words*, 59–60.
87. Harrisson, 'Radio' column.
88. Marie Slocombe, with Tim Eckersley, established a sound library at this time when few were interested in such a pursuit. However, from the BBC's Written Archives, it is evident that there were tensions with Koch about his endless requests during the war to further develop the sound library when resources were limited.
89. Ludwig Koch, 'A Blackbird Mimic', *Times*, 13 May 1940, 4.
90. Mass Observation report, 'Cars and Sirens', 1940, quoted in Peter Adey, 'The Private Life of an Air Raid. Mobility, Stillness, Affect', in *Stillness in a Mobile World*, ed. David Bissell and Gillian Fuller (London: Routledge, 2011), 291.
91. Briggs, *The Golden Age*, 13. Briggs illustrates this point with a poem about the BBC published in *Punch* on 11 May 1932, which begins: '. . . consoling voices of the air / Soothing the sightless, cheering the bedridden'.
92. Koch, *Memoirs of a Birdman*, 71.
93. 'Sound Pictures. Early Morning on a Hampshire Farm', April 1942, British Library, Ludwig Koch gramophone recording number 1LL0003863.
94. Angus Calder, *The People's War: Britain 1939–45* (London: Pimlico, 1992), 318, 366.
95. 'Tweet-Tweet', letter, *Radio Times*, 19 February 1943, 210.
96. Koch, *Memoirs of a Birdman*, 179.
97. Nicholson and Koch, *Songs of Wild Birds*, xiii.
98. Nicholson and Koch, *Songs of Wild Birds*, xiv.
99. Nicholas, 'The People's Radio', 63. Nicholas argues that in wartime, radio came to align itself with 'the people' and therefore the war effort as a whole, thus diminishing interwar social antagonisms of highbrow and lowbrow musical tastes. Variety and popular music increased, and comedy and talks acquired a classless quality in their appeal, with working-class tastes especially catered for: Scannell and Cardiff, *A Social History of British Broadcasting*, 221.

100. Nicholas, *The Echo of War*, 100.

101. Nicholas, *The Echo of War*, 239–40. On ideas of ordinariness in culture, see Jo Fox, 'Millions Like Us? Accented Language and the "Ordinary" in British Films of the Second World War', *The Journal of British Studies*, 45 (2006), 819–45.

102. Attitudes to German people in relation to the Nazi campaign hardened during the war. By 1943, many believed that there was no difference between Germans and Nazis: Nicholas, *The Echo of War*, 160.

103. For more on Koch's style as a broadcaster, see Simon Elmes, *Hello Again: Nine Decades of Radio Voices* (London: Arrow Books, 2012), 192–4.

104. Nicholas, *The Echo of War*, 229.

105. Nicholas, *The Echo of War*, 233.

106. Sonya O. Rose, *Which People's War? National Identity and Citizenship in Wartime Britain 1939–1945* (New York: Oxford University Press, 2004), 213–14, 289.

107. The idea of the rural was usually associated with England, rather than Britain, but the relationship is complex. See Howkins, 'The Discovery of Rural England', 85–111.

108. Frank Newbould's September 1942 recruitment poster campaign *Your Britain, Fight for It Now* depicted pastures and leafy villages of an evocative but bygone era for most people. During WWI the Parliamentary Recruiting Committee had used idyllic rural imagery of southern English landscape in its 1915 campaign.

109. J. B. Priestley, quoted in Angus Calder, *The Myth of the Blitz* (London: Pimlico, 1992), 185.

110. For brief sonic analyses of *Listen to Britain* see Mansell, *The Age of Noise*, 178–80; E. Anna Claydon, 'National Identity, the GPO Film Unit and their Music', 183. Claydon argues that because Britain was still an 'aural media nation' in the 1930s and 40s the use of sound to provoke national pride in *Listen to Britain* was more important than visualizing it.

111. Fisher, *Watching Birds*, 13.

112. Fisher, *Watching Birds*, 11.

113. Moss, *A Bird in the Bush*, 168.

114. 'The Song Thrush Is Silent in August', 18 September 1944, BBC Written Archive, Ludwig Koch scripts.

115. Nicholson and Koch, *Songs of Wild Birds*, 26.

116. Mugglestone, 'Rethinking the Birth of an Expression'. Some evidence suggests that the posters were never displayed at all.

117. 'Starman's Diary', *The Star*, 8 April 1943.

118. R. S. R. Fitter, *London's Natural History* (London: Collins, 1945), 229–30. Starlings continued to roost in London and the notable heronry in Walthamstow was little affected by close bombing, Fitter also reported.

119. There is an expansive literature about animal-human relations and the place of animals in human history. Some works that have informed this section are John Berger, *Why Look at Animals* (London: Penguin, 2009), 12–37; Gray, *The Silence of Animals*, 148–62; Harriet Ritvo, *The Animal Estate: The English and Other Creatures in the Victorian Age* (Cambridge, MA: Harvard University Press, 1989); Mark Cocker, *Birds and People* (London: Jonathan Cape, 2013); Brian Fagan, *The Intimate*

Bond: How Animals Shaped Human History (New York: Bloomsbury, 2015); Claude Lévi-Strauss, *Totemism* (London: Merlin Press, 1964).

120. See, for example, Felipe Fernandez-Armesto, *Civilizations: Culture, Ambition, and the Transformation of Nature* (New York: Touchstone, 2001).
121. Huxley, *Bird-Watching*, 116.
122. Huxley, *Bird-Watching*, 113–14.
123. Bowler, *Science for All*, 46–8.
124. Huxley, *Bird-Watching*, 5.
125. Nicholson and Koch, *Songs of Wild Birds*, 183.
126. Koch, *Memoirs of a Birdman*, 11.
127. Book Chronicle, *The Listener*, 4 November 1936, 877.
128. Koch, *Memoirs of a Birdman*, 42.
129. Koch, *Memoirs of a Birdman*, 113–15. Sir William Haley, director general of the BBC, responded to the recognition UNESCO had given Koch's work by providing the resources for him to process his recordings over the next three years. On Huxley's work with UNESCO see Marianne Sommer, 'Animal Sounds Against the Noise of Modernity and War: Julian Huxley (1887–1975) and the Preservation of the Sonic World Heritage', *Journal of Sonic Studies*, accessed 2 June 2020, https://www.researchcatalogue.net/view/325229/325230.
130. See 'General Conference First Session Held at UNESCO House, Paris, from 20 November to 10 December 1946' (UNESCO, 1947), 222, accessed 2 June 2020, http://unesdoc.unesco.org/images/0011/001145/114580e.pdf.
131. Reynolds, *The Long Shadow*, 162–3. For a contemporary discussion of civilization as a destructive and historical phase in human development to be superseded by a shift towards engagement with the powers of nature, see Edward Carpenter, *Civilisation: Its Cause and Cure and Other Essays* (London George Allen & Unwin, 1921 [1889]).
132. Carpenter, *Civilisation*, 198–207; Overy, *The Morbid Age*.
133. Overy, *The Morbid Age*, 4. Scientific modernity in particular had revealed concerns that included the power of genetic inheritance and the possibility of racial degeneration, the unknown forces of the unconscious that psychologists said may undermine rational thinking, and the unsustainability of the capitalist social system. Mansell has pointed to the continuities between interwar critiques of urban noise that relied on a pathologizing of modernity by medical elites and discourses of modern civilization at risk in Europe and America: 'Neurasthenia, Civilization and the Sounds of Modern Life: Narratives of Nervous Illness in Interwar Campaigns against Noise', in *Sounds of Modern History*, 278–304.
134. Overy, *The Morbid Age*, 3. A counter to the pessimistic orthodoxy of some analyses of cultural and economic gloom in this period is found in Martin Pugh, *We Danced All Night: A Social History of Britain Between the Wars* (London: Vintage, 2009).
135. Reynolds, *The Long Shadow*, 198.
136. Apart from economic planning, Overy highlights a turn to utopian politics of the right or left, moral and religious revival, world government, and eugenic engineering: *The Morbid Age*, 4.

137. Daniel Ritschel, *The Politics of Planning: The Debate on Economic Planning in the Britain in the 1930s* (Oxford: Clarendon Press, 1997), 154–6. See also Arthur Marwick, 'Middle Opinion in the Thirties: Planning, Progress and Political "Agreement," ' *English Historical Review,* 79 (1964), 286–90.

138. Ritschel, *The Politics of Planning,* 144–60; Overy, *The Morbid Age,* 82–3.

139. Overy, *The Morbid Age,* 83. Overy sketches the activities of PEP as a response to the certainty across party-political lines that capitalism needed urgent reform (81–86).

140. Reith, *Into the Wind,* 103.

141. Harman Grisewood, *One Thing at a Time: An Autobiography* (London: Hutchinson, 1968), 129.

142. Quoted in 'The Unspeakable Atrocity', *Document,* BBC Radio 4, 26 August 1993. A recording of the programme is held at the Imperial War Museum, catalogue number 13730. The focus of this programme is accusations of anti-Semitism in the BBC and Foreign Office.

143. Grisewood, *One Thing at a Time,* 130.

144. Tim Crook, *International Radio Journalism: History, Theory and Practice* (London: Routledge, 2002), 196; Humphrey Carpenter, *The Envy of the World: Fifty Years of the Third Programme and Radio Three* (London: Phoenix, 1997), 67; and Curran and Seaton, *Power Without Responsibility,* 125.

145. Crook, *International Radio Journalism,* 196.

146. Quoted in Colls, *Identity of England,* 206.

147. Samuel, *Theatres of Memory,* 219.

148. Brian Currid, *A National Acoustics: Music and Mass Publicity in Weimar and Nazi Germany* (Minneapolis: University of Minnesota Press, 2006).

149. J. B. Priestley, 'Sunday, 23rd June, 1940', *Postscripts* (London: William Heinemann, 1940), 17.

150. Pat Murphy, 'Hitler's Voice', *Daily Mail,* 28 April 1939, 10.

151. Internal BBC communications quoted in Nicholas, *The Echo of War,* 151.

152. John Baxendale, *Priestley's England: J. B. Priestley and English Culture* (Manchester: Manchester University Press, 2007), 140. A third of the country or more listened to the *Postscripts* talks.

153. Priestley, 'Wednesday, 5th June 1940', *Postscripts,* 1–4.

154. Priestley in his *Postscripts* talks from June to September quoted in Baxendale, *Priestley's England,* 150.

155. Priestley, 'Sunday, 9th June, 1940', *Postscripts,* 5.

156. Peter Denney, Bruce Buchan, David Ellison, and Karen Crawley, eds., *Sound, Space and Civility in the British World, 1700–1850* (Abingdon: Routledge, 2020).

157. Priestley, 'Sunday 7th July, 1940', *Postscripts,* 25–6.

158. Priestley, 'Sunday, 9th June, 1940', *Postscripts,* 7.

159. Priestley, 'Sunday, 9th June, 1940', *Postscripts,* 7.

160. Sonia Roberts, *Bird-Keeping and Birdcages: A History* (Newton Abbot: David & Charles, 1972), 67, 73.

Afterword

1. 'Some Thoughts on the Common Toad', in *Smothered Under Journalism: 1946 (The Complete Works of George Orwell)*, ed. Peter Davison (London: Secker & Warburg, 2001), 238–40.
2. Edward Thomas, *The South Country* (Toller Fratrum, Devon: Little Toller, 2009 [1909]), 29.

References

Archive sources

BBC Written Archive Centre, Caversham: Ludwig Koch radio scripts and papers.
Bishopsgate Institute, London: Catherine Gayler diary.
British Red Cross, London: *The Red Cross: The Official Journal of the British Red Cross*; Homes of Recovery papers.
Enham Village Centre archive, Enham Trust, Hampshire.
Imperial War Museum, London: WWI photography collection; WWI recruitment and propaganda poster collection; private papers of Lieutenant J. H. Butlin; Enham Village Centre papers.
Oxford University's First Word War Poetry Digital Archive: *The Hydra* magazine, http://www.oucs.ox.ac.uk/ww1lit/collections/publications.
Private journals and papers kept by the family of Willis Marshall.
Royal College of Music, London: Harrison Sisters' Collection.
Tate Archive, London: Paul Nash papers and letters.
The National Archives, Kew: Ex-Services Welfare Society papers.
Wellcome Library, London: Lieutenant Colonel John P. Silver papers; *War Neurosis. Netley Hospital, 1917*, film, 1918 (Arthur Hurst and J. L. M. Symns).

Sound recordings, radio programmes, and films

All Quiet on the Western Front. Directed by Lewis Milestone, 1930.
'Beatrice Harrison, Cello and Nightingale Duet 19 May 1924', BBC Radio 4. Accessed 30 March 2021, http://www.bbc.co.uk/programmes/p01z12h7.
Bonner, Sam. Audio recording, 1924, British Library catalogue number C653/3.
Burton, John and Desmond Hawkins. *A Salute to Ludwig Koch*. London: BBC Records, 1969, 33 rpm LP.
Listen to Britain. Directed by Humphrey Jennings and Stewart McAllister, 1942.
London Can Take It! Directed by Harry Watt and Humphrey Jennings, 1940.
Nicholson, E. M. and Ludwig Koch. *More Songs of Wild Birds*. London: Witherby, 1937, 78 rpm gramophone disc.
Nicholson, E. M. and Ludwig Koch. *Songs of Wild Birds*. London: Witherby, 1951 (1936), 78 rpm gramophone disc.
Nightingales: A Celebration. Thetford: British Trust for Ornithology, 2009, compact disc.
Street, Seán. *Ludwig Koch and the Music of Nature*. BBC Radio 4, 15 April 2009. Accessed 30 March 2021, https://www.bbc.co.uk/sounds/play/b00jn4m2.
The Demi-Paradise. Directed by Anthony Asquith, 1943.
'The Unspeakable Atrocity', *Document*, BBC Radio 4, 26 August 1993.
Things to Come. Directed by William Cameron Menzies, 1936.

Collections of contemporary material

Beedham, Ann. *Days of Sunshine and Rain: Rambling in the 1920s*. Sheffield: You Books, 2011.

Blunden, Edmund and Martin Taylor, eds. *Overtones of War: Poems of the First World War*. London: Duckworth, 1996.

Davison, Peter, ed. *Orwell's England*. London: Penguin, 2001.

Davison, Peter, ed. *Smothered Under Journalism: 1946 (The Complete Works of George Orwell)*. London: Secker & Warburg, 2001.

Day-Lewis, Cecil, ed. *The Collected Poems of Wilfred Owen*. New York: New Directions Books, 1965.

Housman, Lawrence. *War Letters of Fallen Englishmen*. New York: E. P. Dutton, 1930.

Kendall, Tim. *Poetry of the First World War: An Anthology*. Oxford: Oxford University Press, 2013.

Laffin, John. *Letters from the Front, 1914–18*. London: J. M. Dent, 1973.

Noakes, Vivien. *Voices of Silence: The Alternative Book of First World War Poetry*. Stroud: The History Press, 2006.

Pollard, Ernest and Hazel Strouts, eds. *Wings over the Western Front: The First World War Diaries of Collingwood Ingram*. Charlbury: Day Books, 2014.

Scollard, C. and J. B. Rittenhouse, eds. *The Bird-Lovers' Anthology*. Boston, MA: Houghton Mifflin, 1930.

Thomas, R. George, ed. *Edward Thomas: Collected Poems and War Diary, 1917*. London: Faber & Faber, 2004.

Webb, Mike. *From Downing Street to the Trenches: First Hand Accounts, from the Great War, 1914–1916*. Oxford: Bodleian, 2014.

Reports, manuals, catalogues

Defence of the Realm Manual, 6th edition. London: HMSO, 1918.

'General Conference First Session Held at UNESCO House, Paris, from 20 November to 10 December 1946'. UNESCO, 1947, 222. Accessed August 10, 2017, http://unesdoc.unesco.org/images/0011/001145/114580e.pdf.

His Master's Voice General Catalogue. London: Gramophone Company, 1911.

His Master's Voice General Catalogue. London: Gramophone Company, 1914.

Report of the War Office Committee of Enquiry into 'Shell-Shock'. London: HMSO, 1922.

Report of the War Office Committee of Enquiry into 'Shell-Shock'. London: Imperial War Museum Military Handbook Series, 2004.

Contemporary newspapers, magazines and journals

'A Kensington Hospital for Officers', *Times*, 18 January 1915, 4.

'A Nest of Singing Birds', *Radio Times*, 13 November 1936, 72.

'A Rural Life for Health and Restoration', *Fruit Grower*, 29 January 1920, 187.

'A Sea Shell Loud Speaker', *Popular Wireless*, 24 November 1923, 471.

'A Word for the Nightingale', *Birmingham Daily Mail*, 24 May 1924, no pagination.

'Bird's Song on Battlefield', *Daily Mail*, 27 April 1915, 3.

'Bren Gunners on the Alert Against Air Attack', *The Listener*, 23 May 1940.

'Come, Birdy, Come!' *Popular Wireless*, 3 June 1922, 7.

'Country Air for Shell-Shock', *Daily Mail*, 26 November 1917, 5.

'Cure by Spring Hues', *Daily Mail*, 27 September 1917, 3.

'Disabled Soldiers in Civil Hospitals', *Times*, 23 January 1918, 3.

'English Traits. Mr Baldwin's Review', *Times*, 7 May 1924, 16.

'Escape from Squalor: Rambling the Best Way', *Manchester Guardian*, 19 June 1933, 13.

'For Shell-Shock Cases', *Times*, 5 November 1918, 3.

'Hikers Miss the Sun. Vain All Night Ramble to Sussex Beacon', *Portsmouth Evening News*, 12 September 1932, 3.

'Hiking Heroism in Yorkshire', *Leeds Mercury*, 24 November 1931, 6.

'Listen to the Wild Waves', *Picture Post*, 9 June 1945, 24.

'Listening-in', *Bristol Times and Mirror*, 22 May 1924, 6.

'Primitive Agents in Treatment', *Lancet*, 21 December 1918, 852.

'Radio in the Summertime', *Radio Times*, 15 May 1925, 343.

'Radio Rest Cures', *Popular Wireless*, 19 April 1924, 262.

'Rambling an Art: Stemming the Tide of Mechanicalism', *Manchester Guardian*, 2 October 1933, 2.

'Rambling as a Fashion', *Manchester Guardian*, 1 December 1936, 13.

'Recalled to Life', *Spectator*, 29 June 1917, 7.

'Rest Gardens tor The Wounded', *Times*, 24 May 1916, 11.

'Shell-Shock Men', *Daily Mail*, 8 November 1917, 5.

'Singing Sick Men to Health. Miracles of Vocal Therapy', *Daily Graphic*, 19 July 1922, 5.

'Soldier and Song Bird', *Daily Mail*, 21 January 1916, 3.

'Starman's Diary', *The Star*, 8 April 1943, no pagination.

'The Care of the Wounded', *British Journal of Nursing*, 5 February 1916, 120.

'The Enham Village Centre for the Re-Education of Men Disabled in the War', *Lancet*, 9 August 1919, 252.

'The Great Years of Their Lives', *The Listener*, 15 July 1971, 74.

'The Happy Warrior', *Reveille*, 1 August 1918, 105.

'The Microphone', *Spectator*, 25 May 1878, 10.

'The Opening of Enham Village Centre', *British Medical Journal*, 8 November 1919, 610.

'The Problem of Nervous Breakdown', *The Spectator*, 13 March 1920, 20.

'The Vocal Therapy Fund', *British Journal of Nursing*, 12 July 1919, 22.

'Tweet-Tweet', letter, *Radio Times*, 19 February 1943, 210.

'When Will Insects Broadcast? The Ultra-Microphone and Its Wonders', *Radio Times*, 14 March 1924, 442.

'Wireless and Health. How Listening Affects Your Well-Being', *Radio Times*, 11 January 1924, 82.

Ash, Edward C. 'Broadcasting Wild Fowl at Night', *Radio Times*, 23 January 1925, 197–8.

Ball, H. Standish. 'The Work of the Miner on the Western Front, 1915–18', *Transactions of the Institution of Mining and Metallurgy*, 28 (1919): 189–248.

Blunt, John. 'Nightingales and Headphones', *Daily Mail*, 20 May 1924, 7.

Book Chronicle, *The Listener*, 4 November 1936, 877.

Book review, *British Birds*, 1 December 1 (1937): 239.

Bruce-Porter, Bruce. 'Health and Headphones', *Radio Times*, 31 July 1925, 1.

Burrows, Arthur. 'Broadcasting the Nightingale', *Radio Times*, 14 December 1923, 428.

By Civilian, 'The Return of the Officer', *Reveille*, 1 (1918): 78–86.

Collie, John. 'The Management of Neurasthenia and Allied Disorders, Contracted in the Army', *Recalled to Life*, September (1917): 234–53.

Corrigan, J. F. 'Aerials in Miniature', *Wireless World*, 1 May 1926, 385–6.

Craven, Arthur B. 'The Yorkshire Ramblers' Club, 1892–1992', *Yorkshire Ramblers' Club Journal*, 11 (1992): 1–5.

de Poynton, E. 'A Mystery of the Animal World. Nature's Own Wireless', *Radio Times*, 12 December 1924, 526.

Draper, Warwick. 'Village Centres for Cure and Training', *Recalled to Life*, 21 April 1918, 342–57.

Eckersley, Peter. 'Broadcasting the Zoo', *Radio Times*, 22 August 1924, 374.

Editorial, 'Broadcasting Birds', *Times*, 21 May 1924, 15.

Elwes, Hervey. 'Five Years' Broadcasting', *Vox*, 11 January 1930, 334.

Ferrars, Lewis J. 'How We Wireless', *The Broadcaster*, June 1923, 46–9.

Fisher, Claude. 'And Now for the Open Road!', *Daily Mail*, 21 March 1932, 10.

Fisher, Claude. 'All the Year Is Hiking Time', *Daily Mail*, 15 September 1932, 4.

Fisher, Claude. 'Take Your Holidays by Moonlight', *Daily Mail*, 18 August 1933, 6.

Fleming, J. A. 'The Polite Use of the Ether', *Radio Times*, 2 July 1926, 41.

Forsythe, D. 'Functional Nerve Disease and the Shock of Battle', *Lancet*, 25 December (1915): 1399–403.

Fox, Fortescue. 'Village Centres', *Reveille*, 1 February 1919, 457–71.

Galsworthy, John. 'Heritage: An Impression', *Reveille*, 1 November 1918, 302–5.

Galsworthy, John. 'The Gist of the Matter', *Reveille*, 1 August 1918, 3–15.

Harding, Major C. Reginald. 'Land Settlement and the Disabled', *Reveille*, 1 February 1919, 457–71.

Harford, Frederick Kill. 'Music in Illness', *Nordic Journal of Music Therapy*, 11 (2002 [1898]): 43–4.

Harrisson, Tom. 'Radio' column, *Observer*, 3 January 1943, 2.

Haydon, Arthur. 'Homes for Shell-Shock Cases', *Daily Mail*, 29 March 1916, 4.

Hicks, Seymour. 'Shell-Shock Cases in Air Raids', *Daily Mail*, 27 December 1917, 2.

Hurst, A. F. 'Cinematograph Demonstration of War Neuroses', *Proceedings of the Royal Society of Medicine*, 11 (1918): 39.

Hurst, A. F. and J. L. M. Symns, 'The Rapid Cure of Hysterical Symptoms in Soldiers', Lancet, 2, 6 August 1918, 140.

Huxley, Aldous. 'The Outlook for American Culture: Some Reflections in a Machine Age', *Harper's Magazine*, August 1927, 265–70.

Huxley, Julian. 'Bird-Watching and Biological Science (Part 1)', *Auk*, 33 (1916): 142–61.

Knutsford, Lord (Sydney Holland). 'Lord Knutsford's Appeal', *Times*, 13 November 1914, 9.

Knutsford, Lord (Sydney Holland). 'Wireless for the Wards', *Radio Times*, 5 June 1925, 504.

Koch, Ludwig. 'A Blackbird Mimic', *Times*, 13 May 1940, 4.

Letter, 'Does Listening Promote Health?' *Radio Times*, 23 November 1923, 320.

Letter. *John O' London's Weekly*, 11 August 1923, 625.

Letter, 'Anti-Desecration Crusade', *Sheffield Daily Telegraph*, 25 May 1928, 3.

Letter, 'Ramblers' Conduct', *Sheffield Daily Telegraph*, 25 May 1928, 3.

Lumsden, Thomas. 'Nerve-Shattered Pensioners', *Times*, 22 August 1917, 9.

Lumsden, Thomas. 'Nerve-Shattered Pensioners', *Times*, 27 August 1917, 8.

Lumsden, Thomas. Letter, *British Medical Journal*, 21 January 1920, 131.

Mallinson, Russell. 'Radio and the Broad Highway', *The Broadcaster*, June 1923, 24.

Mallinson, Russell. 'The Sunny Side of Radio', *The Broadcaster*, August 1923, 18.

Medlicott, W. S. 'Bird Notes from the Western Front (Pas-de-Calais)', *British Birds*, 12 (1919): 271–5.

Milner, Frederick. Letter, *John O'Groat Journal*, 14 October 1921, 3.

Mitchison, Mary E. 'A Debt to the Wounded', *Times*, 23 May 1916, 9.

Murphy, Pat. 'Hitler's Voice', *Daily Mail*, 28 April 1939, 10.

Myers, C. S. 'A Contribution to the Study of Shell Shock: Being an Account of Three Cases of Loss of Memory, Vision, Smell, and Taste, Admitted into the Duchess of Westminster's War Hospital, Le Touquet', *Lancet*, 13 February 1915, 316–20.

Noyes, Alfred. 'Radio and the Master-Secret', *Radio Times*, 18 September 1925, 549–50.

Pam, Albert. 'Bird Life on the Battlefields', *The Avicultural Magazine*, 1917, 239.

Pick, Push. 'V. C. Canaries', *Daily Mail*, 12 September 1918, 2.

Preece, W. H. 'The Phonograph', *Journal of the Society of the Arts*, 26 (1878): 534–8.

Reith, John. 'Concerning Tinned Nightingale', *Radio Times*, 11 April 1924, 85–6.

Reith, John. 'The Broadcasting of Silence', *Radio Times*, 6 June 1924, 437–8.

Reith, John. 'The Lure and Fear of Broadcasting. A Reply to John O' London', *John O' London's Weekly*, 22 March 1924, 938.

Reith, John. 'When Silence Was Served', *Radio Times*, 12 June 1925, 529–30.

Risdon, P. J. 'Nature's Wireless', *Popular Wireless*, 22 March 1924, 123.

Rivers, W. H. R. 'An Address on the Repression of War Experience', *Lancet*, 2 February 1918, 173–7.

Russell, Sinclair. 'The Radio Pipes of Pan', *The Broadcaster*, July 1923, 58–60.

Scott, R. Hamilton. 'Birds in and Around the Firing Line', *The Avicultural Magazine*, 1918, 247–8.

Selfridge, Gordon. 'Mr Selfridge Expresses His Views', *Popular Wireless*, 3 June 1922, 7.

Walshe, Elyn. 'Green Roads: Solitude and the Litter Fiends', *Manchester Guardian*, 1 May 1933, 6.

Whitten, Wilfred. 'A Word About Mr Broadcast', *John O' London's Weekly*, 13 December 1924, 433.

Whitten, Wilfred. 'Pray Silence', *John O' London's Weekly*, 21 July 1923, 517–18.

Whitten, Wilfred. 'The Lure and Fear of Broadcasting', *John O' London's Weekly*, 15 March 1924, 865.

Wills, Frederick Alexander. *The Rambles of 'Vagabond' of the 'Newcastle Evening Chronicle'*, Newcastle Upon Tyne: J & P Bealls, 1936.

Woolf, Virginia. 'Thoughts on Peace in an Air Raid', *New Republic*, 21 October 1940, 549–51.

Contemporary books

Ash, Edwin. *The Problem of Nervous Breakdown*. New York: Macmillan, 1920.

Beard, George. *American Nervousness: Its Causes and Consequences*. New York: G. P. Putnam's, 1881.

Blunden, Edmund. *Undertones of War*. London: Penguin, 2000 (1928).

Brooks, Edgar. *Country Rambles Round Birmingham and Week-End Holidays for the Pedestrian, Cyclist, and Motorist*, 6th edn. Birmingham: Cornish Brothers, 1927.

Burrows, Arthur. *The Story of Broadcasting*. London: Cassell, 1924.

Childs, W. M. *Holidays in Tents*. London: J. M. Dent, 1921.

Fisher, James. *Watching Birds*. London: Penguin, 1946 (1941).

Fitter, R. S. R. *London's Natural History*. London: Collins, 1945.

Garstang, Walter. *Songs of the Birds*. London: John Lane The Bodley Head, 1922.

Gibbs, Philip. *Now It Can be Told*. New York: Harper and Brothers, 1920.

Gillespie, A. D. *Letters from Flanders*. London: Smith, Elder, 1916.

Gladstone, Hugh. *Birds and the War*. London: Skeffington, 1919.

Graham, Stephen. *The Gentle Art of Tramping*. London: Bloomsbury, 2019 (1927).

Graves, Robert. *Goodbye to All That*. London: Penguin, 1960 (1929).

Graves, Robert and Alan Hodge. *The Long Week-End: A Social History of Great Britain 1918–1939*. Harmondsworth: Penguin, 1971 (1940).

Grey, Edward. *Fallodon Papers*. Boston, MA: Houghton Mifflin, 1926.

Grey, Edward. *The Charm of Birds*. London: Hodder and Stoughton, 1927.

Hargrave, John. *The Great War Brings It Home: The Natural Reconstruction of an Unnatural Existence*. London: Constable, 1919.

Harrisson, Tom and Charles Madge. *War Begins at Home*. London: Chatto & Windus, 1940.

Harris, Wilfred. *Nerve Injuries and Shock*. London: Henry Frowde, 1915.

Haultain, Arnold. *Of Walking and Walking Tours: An Attempt to Find a Philosophy and Creed*. London: T. Werner Laurie, 1914.

Holiday Songs. London: The Holiday Fellowship, 1937.

Humble, B. H. and W. M. McLellan. *Songs for Climbers*. Glasgow: W. M. McLellan, 1938.

Huxley, Julian. *Bird-Watching and Bird Behaviour*. London: Chatto & Windus, 1930.

Joad, C. E. M. *The Horrors of the Countryside*. London: Hogarth Press, 1931.

Joad, C. E. M. *A Charter for Ramblers: The Future of the Countryside*. London: Hutchinson, 1934.

Joad, C. E. M. *The Untutored Townsman's Invasion of the Countryside*. London: Faber and Faber, 1946.

Jünger, Ernst. *Storm of Steel*. London: Penguin, 2004 (1920).

Kerr, Wilfred. *Shrieks and Crashes: Being Memories of Canada's Corps, 1917*. Toronto: H. Rose, 1929.

Knutsford, Lord (Sydney Holland). *In Black and White*. London: Edward Arnold, 1927.

Koch, Ludwig. *Memoirs of a Birdman*. London: Phoenix House, 1955.

Laborde, E. D. *Popular Map Reading*. Cambridge: Cambridge University Press, 1928.

Lodge, Oliver. *Ether and Reality: A Series of Discourses on the Many Functions of the Ether of Space*. London: Hodder & Stoughton, 1925.

Lodge, Oliver. *Talks About Wireless: With Some Pioneering History and Some Hints and Calculations for Wireless Amateurs*. London: Cassell, 1926.

McKenzie, Dan. *Aromatics of the Soul: A Study of Smells*. London: Heinemann, 1923.

McKenzie, Dan. *City of Din: A Tirade Against Noise*. London: Adlard, 1916.

Millicent Ashdown, Amy. *A Complete System of Nursing*. London: J. M. Dent, 1917.

Moorhouse, Sydney. *Walking Tours and Hostels in England*. London: Country Life, 1936.

Morris, William. *News from Nowhere and Other Writings*, edited by Clive Wilmer. London: Penguin Classics, 1993 (1890).

Morton, H. V. *H. V. Morton's London*. London: Methuen, 1948 (1942).

Mott, Frederick W. *War Neuroses and Shell Shock*. London: Oxford University Press, 1919.

Munro, H. H. *The Square Egg and Other Sketches*. London: John Lane The Bodley Head, 1924.

Murray, Geoffrey. *The Gentle Art of Walking*. London: Blackie, 1939.

Nash, Paul. *Aerial Flowers*. Oxford: Counterpoint, 1947.

Nicholson, E. M. *The Art of Birdwatching: A Practical Guide to Field Observation*. London: Witherby, 1931.

Nicholson, E. M. and Ludwig Koch. *More Songs of Wild Birds*. London: Witherby, 1937.

Nicholson, E. M. and Ludwig Koch. *Songs of Wild Birds*. London: Witherby, 1951 (1936).

Nightingale, Florence. *Notes on Nursing: What It Is and What It Is Not*. Philadelphia: Stern, 1946 (1859).

Philipps, Colwyn. *Verses*. London: Smith, Elder, 1915.

Priestley, J. B. *Postscripts*. London: William Heinemann, 1940.

Pulbrook, Ernest. *The English Countryside*. London: B. T. Batsford, 1915.

Raven, Charles. *In Praise of Birds*. London: Martin Hopkinson, 1925.

Reith, John. *Broadcast over Britain*. London: Hodder & Stoughton, 1924.

Reith, John. *Into the Wind*. London: Hodder & Stoughton, 1949.

Reith, John. *Wearing Spurs*. London: Hutchinson, 1966.

Remarque, Erich Maria. *All Quiet on the Western Front*. London: Vintage, 1996 (1929).

Robinson, E. Kay. *At Home with Nature*. London: Hodder and Stoughton, 1924.

Robinson, E. Kay. *Religion of Nature*. London: Hodder and Stoughton, 1906.

Robinson, E. Kay. *The Country Day by Day*. London: William Heinemann, 1905.

Robinson, E. Kay. *The Meaning of Life*. London: Hampton Wick, 1916.

Saville-Sneath, R. A. *Aircraft Recognition*. London: Penguin, 1990 (1941).

Shepherd, Nan. *The Living Mountain*. London: Canongate, 2014 (1977).

Sidgwick, Arthur. *Walking Essays*. London: Edward Arnold, 1912.

Sims, George, ed. *Living London*, Volume 3, Section 1. London: Cassell, 1901.

Sinclair, W. A. *The Voice of the Nazi: Being Eight Broadcast Talks Given Between December 1939 and May 1940*. London, Collins, 1940.

Smith, Percy. *Sixteen Drypoints and Etchings: A Record of the Great War*. London: Soncino Press, 1930.

Stephen, Leslie. *Studies of a Biographer*. London: Duckworth, 1902.

Stephenson, Tom. *The Countryside Companion*. London: Odhams, 1939.

Stevenson, Robert Louis. *The Works of Robert Louis Stevenson*, Miscellanies, Volume III. London: Chatto & Windus, 1895.

The Hackney Scout Song Book. London: Stacy & Son, 1921.

Thomas, Edward. *The South Country*. London: J. M. Dent, 1909.

Thomas, Edward. *The South Country*. Toller Fratrum, Devon: Little Toller, 2009 (1909).

Thurstan, Violetta. *A Text Book of War Nursing*. London: G. P. Putnam, 1917.

Trevelyan, G. M. *Clio, A Muse and Other Essays Literary and Pedestrian*. London: Longmans, Green, 1913.

Turner, Emma. *Every Garden a Bird Sanctuary*. London: Witherby, 1935.

Vaughan, Edwin Campion. *Some Desperate Glory: The Diary of a Young Officer 1917*. Barnsley: Pen & Sword, 2010.

Ward, G. H. B. *Songs for Ramblers to Sing on the Moorlands*. Sheffield: Sheffield Clarion Ramblers, 1922.

West, Arthur Graeme. *The Diary of a Dead Officer, Being the Posthumous Papers of Arthur Graeme West*. London: Allen and Unwin, 1919.

Whitty, H. Ramsden. *Observers' Tale: The Story of Group 17 of the R.O.C.* London: Roland Brothers, 1951.

Williams-Ellis, Clough. *Britain and the Beast*. London: J. M. Dent, 1937.

Williams-Ellis, Clough. *England and the Octopus*. London: Geoffrey Bles, 1929.

Williamson, Henry. *Tarka the Otter*. Harmondsworth: Penguin, 1961 (1927).

Secondary books and journals

Abram, David. *The Spell of the Sensuous: Perception and Language in a More-Than-Human World*. New York: Vintage, 1997.

Aldgate, Tony and Jeffrey Richards. *Britain Can Take It: British Cinema in the Second World War*. London: I. B. Tauris, 2007.

Allen, David Elliston. *The Naturalist in Britain: A Social History*. Princeton, NJ: Princeton University Press, 1994.

Anderson, Ben. 'A liberal countryside? The Manchester Ramblers' Federation and the "social readjustment" of urban citizens, 1929–1936', *Urban History*, 38 (2011), 84–102.

Anthony, Scott and James G. Mansell. *The Projection of Britain: A History of the GPO Film Unit*. Basingstoke: Palgrave Macmillan, 2011.

Armitage, Simon and Tim Dee. *The Poetry of Birds*. London: Penguin, 2011.

Arnold, Matthew. *Culture and Anarchy and Other Writings*, ed. Stefan Collini. Cambridge: Cambridge University Press, 2002.

Attali, Jacques. *Noise: The Political Economy of Music*. Minneapolis: University of Minnesota Press, 1985.

Avery, Todd. *Radio Modernism: Literature, Ethics, and the BBC, 1922–1938*. Aldershot: Ashgate, 2006.

Baade, Christina. *Victory Through Harmony: The BBC and Popular Music in World War II*. New York: Oxford University Press, 2012.

Bailey, Michael, ed. *Narrating Media History*. London: Routledge, 2009.

Bailey, Peter. *Leisure and Class in Victorian England: Rational Recreation and the Contest for Control, 1830–1885*. London: Routledge, 2007 (1978).

Bailey, Peter. 'Breaking the Sound Barrier: A Historian Listens to Noise', *Body and Society*, 2 (1996), 49–66.

Barham, Peter. *Forgotten Lunatics of the Great War*. New Haven, CT: Yale University Press, 2007.

Barker, Ralph. *The Royal Flying Corps in World War One*. London: Robinson, 2002.

Barrie, Alexander. *War Underground*. London: Star, 1981.

Bates, H. E. *Flying Bombs over England*, large print edition. Oxford: ISIS, 1997.

Baxendale, John. *Priestley's England: J. B. Priestley and English Culture*. Manchester: Manchester University Press, 2007.

Beard, Mary. 'The Public Voice of Women', *London Review of Books*, 20 March 2014, 11–14.

Ben-Ze'ev, E., R. Genie, and J. Winter, eds. *Shadows of War: A Social History of Silence in the Twentieth Century*. Cambridge: Cambridge University Press, 2010.

Bennett, Tony, Colin Mercer and Janet Woollacott, eds. *Popular Culture and Social Relations*. Milton Keynes: Open University Press, 1986.

Berger, John. *Why Look at Animals*. London: Penguin, 2009.

Bhatti, Mark, Andrew Church, Amanda Claremont and Paul Stenner. '"I Love Being in the Garden": Enchanting Encounters in Everyday Life', *Social & Cultural Geography*, 10 (2009), 61–76.

Bialer, Uri. *The Shadow of the Bomber: The Fear of Air Attack and British Politics, 1932–1939*. London: Royal Historical Society, 1980.

Biddle, Ian and Kirsten Gibson. *Cultural Histories of Noise, Sound and Listening in Europe, 1300–1918*. London: Routledge, 2016.

Bijsterveld, Karin. *Mechanical Sound: Technology, Culture and Public Problems of Noise in the Twentieth Century*. Cambridge, MA: MIT Press, 2008.

Bissell, David and Gillian Fuller. *Stillness in a Mobile World*. London: Routledge, 2011.

Bluemel, Kirstin and Michael McCluskey, *Rural Modernity in Britain: A Critical Intervention*. Edinburgh: Edinburgh University Press, 2018.

Bogacz, Ted. 'War Neurosis and Cultural Change in England, 1914–22: The Work of the War Office Committee of Enquiry into "Shell-Shock"', *Journal of Contemporary History*, 24 (1989), 227–56.

Bourke, Joanna. *Dismembering the Male: Men's Bodies, Britain and the Great War*. London: Reaktion, 1999.

Bourke, Joanna. *Fear: A Cultural History*. London: Virago Press, 2006.

Bowler, Peter. *Science for All: The Popularisation of Science in Early Twentieth-Century Britain*. Chicago: University of Chicago Press, 2009.

Boyd Haycock, David. *Paul Nash*. London: Tate Publishing, 2002.

Boym, Svetlana. *The Future of Nostalgia*. New York: Basic Books, 2001.

Brassley, Paul, Jeremy Burchardt, and Lynne Thompson. *The English Countryside Between the Wars: Regeneration or Decline?* Woodbridge: The Boydell Press, 2006.

Bright, Michael. *100 Years of Wildlife*. London: BBC Books, 2007.

Briggs, Asa. *The History of Broadcasting in the United Kingdom*, vol. 1, *The Birth of Broadcasting*. Oxford: Oxford University Press, 2000.

Briggs, Asa. *The History of Broadcasting in the United Kingdom*, vol. 2, *The Golden Age of Wireless*. London: Oxford University Press, 1965.

Briggs, Asa. *The History of Broadcasting in the United Kingdom*, vol. 3, *The War of Words*. Oxford: Oxford University Press, 1995.

Brooks, Jane and Christine Hallett, eds. *One Hundred Years of Wartime Nursing Practices, 1854–1953*. Manchester: Manchester University Press, 2015.

Bruton, Elizabeth and Graeme Gooday. 'Listening in Combat: Surveillance Technologies Beyond the Visual in the First World War', *History and Technology*, 32 (2016), 213–26.

Bruyninckx, Joeri. *Listening in the Field: Recording and the Science of Birdsong*. Cambridge, MA: MIT Press, 2018.

Bryder, Linda. 'Papworth Village Settlement: A Unique Experiment in the Treatment and Care of the Tuberculosis?', *Medical History*, 28 (1984), 372–90.

Bunn, Stephanie, ed. *Anthropology and Beauty: From Aesthetics to Creativity*. Abingdon: Routledge, 2020.

Burchardt, Jeremy. *Paradise Lost: Rural Idyll and Social Change Since 1800*. London: I. B. Tauris, 2002.

Burchardt, Jeremy. 'Rurality, Modernity and National Identity Between the Wars', *Rural History*, 21 (2010), 143–50.

Burton, John. 'The BBC Natural History Unit Wildlife Sound Library 1948–1988', *Wildlife Sound*, 12 (2012), 19.

Butler, Shane, and Sarah Nooter, eds. *Sound and the Ancient Senses*. New York: Routledge, 2018.

Bynum, Helen. *Spitting Blood: The History of Tuberculosis*. Oxford: Oxford University Press, 2012.

Calder, Angus. *The Myth of the Blitz*. London: Pimlico, 1992.

Calder, Angus. *The People's War: Britain 1939–45*. London: Pimlico, 1992.

Carey, John. *The Intellectuals and the Masses: Pride and Prejudice Among the Literary Intelligentsia 1880–1939*. London: Faber and Faber, 1992.

Carpenter, Humphrey. *The Envy of the World: Fifty Years of the Third Programme and Radio Three*. London: Phoenix, 1997.

Chantler, A. and R. Hawkes, eds. *Ford Madox Ford's Parade's End: The First World War, Culture, and Modernity*. Amsterdam: Rodopi, 2014.

Classen, Constance. 'Engendering Perception: Gender Ideologies and Sensory Hierarchies in Western History', *Body & Society*, 3 (1997), 1–19.

Classen, Constance, ed. *The Book of Touch*. Abingdon: Routledge, 2020.

Clayton, Martin, Rebecca Sager and Udo Will. 'In Time with the Music: The Concept of Entrainment and Its Significance for Ethnomusicology', *European Seminar in Ethnomusicology Counterpoint*, 1 (2005), 1–82.

Cleveland-Peck, Patricia. *The Cello and the Nightingales: The Autobiography of Beatrice Harrison*. London: John Murray, 1985.

Cloete, Stuart. *A Victorian Son: An Autobiography, 1897–1922*. London: Collins, 1972.

Cockayne, Emily. *Hubbub, Filth, Noise & Stench in England 1600–1770*. New Haven, CT: Yale University Press, 2008.

Cocker, Mark. *Birds and People*. London: Jonathan Cape, 2013.

Cocker, Mark and Richard Mabey. *Birds Britannica*. London: Chatto & Windus, 2005.

Colls, Robert and Philip Dodd, eds. *Englishness: Politics and Culture 1880–1920*. London: Bloomsbury, 2014.

Colls, Robert. *Identity of England*. Oxford: Oxford University Press, 2002.

Cooter, R., M. Harrison and S. Sturdy, eds. *War, Medicine and Modernity*. Stroud: Sutton, 1999.

Copeland, Peter, Jeffrey Boswall, and Leonard Petts. *Birdsongs on Old Records: A Coarsegroove Discography of Palearctic Region Bird Sound 1910–1958*. London: British Library Wildlife Sounds, National Sound Archive, 1988.

Cowgill, Rachel. 'Canonizing Remembrance: Music for Armistice Day at the BBC, 1922–7', *First World War Studies*, 2 (2011) 75–107.

Craiglockhart, History of. War Poets Collection at Edinburgh Napier University. Accessed 3 September 2017, http://www2.napier.ac.uk/warpoets/1800.htm#1800.

Crook, Tim. *International Radio Journalism: History, Theory and Practice*. London: Routledge, 2002.

Crouthamel, Jason and Peter Leese. *Psychological Trauma and the Legacies of the First World War*. Basingstoke: Palgrave Macmillan, 2016.

Curran, James and Jean Seaton. *Power Without Responsibility: Press and Broadcasting in Britain*. London: Routledge, 1997.

Currid, Brian. *A National Acoustics: Music and Mass Publicity in Weimar and Nazi Germany*. Minneapolis: University of Minnesota Press, 2006.

Dakers, Caroline. *The Countryside at War*. London: Constable, 1987.

Daniel, Ute, Peter Gatrell, Oliver Janz, Heather Jones, Jennifer Keene, Alan Kramer, and Bill Nasson, eds. *International Encyclopaedia of the First World War*. Issued by Freie Universität Berlin, Berlin, 8 October 2014. Accessed 5 January 2017, http://dx.doi.org/10.15463/ie1418.10453.

Darley, Gillian. *Villages of Vision: A Study of Strange Utopias*. Nottingham: Five Leaves, 2007.

Das, Santanu. *Touch and Intimacy in First World War Literature*. Cambridge: Cambridge University Press, 2005.

Daston, Lorraine and Peter Galison. *Objectivity*. Brooklyn, NY: Zone Books, 2007.

Daughtry, J. Martin. *Listening to War: Sound, Music, Trauma and Survival in Wartime Iraq*. New York: Oxford University Press, 2015.

Davison, Peter, ed. *Smothered Under Journalism: 1946 (The Complete Works of George Orwell)*. London: Secker & Warburg, 2001.

Denney, Peter, Bruce Buchan, David Ellison, and Karen Crawley, eds. *Sound, Space and Civility in the British World, 1700–1850*. Abingdon: Routledge, 2020.

DeGroot, Gerard J. *Blighty: British Society in the Era of the Great War*. Harlow: Pearson, 1996.

Douglas, Mary. *Purity and Danger: An Analysis of Concept of Pollution and Taboo*. London: Routledge, 2002.

Douglas, Susan. *Listening in: Radio and the American Imagination*. Minneapolis: University of Minnesota Press, 2004.

Dunn, Melanie. 'Hysterical War Neuroses: A Study of Seale-Hayne Neurological Military Hospital, Newton Abbot, 1918–1919', MA diss., University of Exeter, 2009.

Edensor, Tim. 'Walking in the British Countryside: Reflexivity, Embodied Practices and Ways to Escape', *Body & Society*, 6 (2000), 81–106.

Edensor, Tim, ed. *Geographies of Rhythm: Nature, Place, Mobilities and Bodies*. Farnham: Ashgate, 2010.

Elmes, Simon. *Hello Again: Nine Decades of Radio Voices*. London: Arrow Books, 2012.

Encke, Julia. 'War Noises on the Battlefield: On Fighting Underground and Learning to Listen in the Great War', *German Historical Institute London Bulletin*, 37 (2015), 7–21.

Enns, Anthony. 'Psychic Radio: Sound Technologies, Ether Bodies, and Vibrations of the Soul', *The Senses and Society*, 3 (2008), 137–52.

Fagan, Brian. *The Intimate Bond: How Animals Shaped Human History*. New York: Bloomsbury, 2015.

Faulks, Sebastian. *Birdsong*. London: Vintage, 1993.

Fauvel, J., R. Flood, M. Shortland, and R. Williams, eds. *Let Newton Be!* Oxford: Oxford University Press, 1988.

Feiereisen, Florence and Alexandra Merley Hill, eds. *Germany in the Loud Twentieth Century: An Introduction*. New York: Oxford University Press, 2012.

Fernandez-Armesto, Felipe. *Civilizations: Culture, Ambition, and the Transformation of Nature*. New York: Touchstone, 2001.

Foster Damon, Samuel. *A Blake Dictionary: The Ideas and Symbols of William Blake*. Hanover, NH: University Press of New England, 1988.

Fox, Jo. 'Careless Talk: Tensions within British Domestic Propaganda during the Second World War', *The Journal of British Studies*, 51 (2012), 936–66.

Fox, Jo. 'Millions Like Us? Accented Language and the "Ordinary" in British Films of the Second World War', *The Journal of British Studies*, 45 (2006), 819–45.

Fussell, Paul. *The Great War and Modern Memory*. New York: Oxford University Press, 1975.

Gardiner, Juliet. *The Animals' War: Animals in Wartime from the First World War to the Present Day*. London: Portrait, 2006.

Geddes, Keith. *The Setmakers: A History of the Radio and Television Industry*. London: BREMA, 1991.

Gijswijt-Hofstra, Marijke and Roy Porter, eds. *Cultures of Neurasthenia from Beard to the First World War*. New York: Rodopi, 2001.

Goddard, Michael, Benjamin Halligan, and Paul Hegarty. *Reverberations: The Philosophy, Aesthetics and Politics of Noise*. London: Continuum, 2012.

Goodale, Greg. *Sonic Persuasion: Reading Sound in the Recorded Age*. Urbana: University of Illinois Press, 2011.

Goodman, Steve. *Sonic Warfare: Sound, Affect, and the Ecology of Fear*. Cambridge, MA: MIT Press, 2010.

Gray, John. *The Silence of Animals: On Progress and Other Modern Myths*. London: Penguin, 2014.

Gregory, Adrian. *The Silence of Memory: Armistice Day, 1919–1946*. Oxford: Berg, 1994.

Griffiths, Clare V. J. *Labour and the Countryside: The Politics of Rural Britain 1918–1939*. Oxford: Oxford University Press, 2007.

Grisewood, Harman. *One Thing at a Time: An Autobiography*. London: Hutchinson, 1968.

Gros, Frédéric. *A Philosophy of Walking*. London: Verso, 2014.

Guida, Michael. '1928. Popular Bird-Watching Becomes Scientific: The First National Bird Census in Britain', *Public Understanding of Science*, 8 (2019), 622–7.

Guida, Michael. 'Nature's Sonic Order on the Western Front', *Transposition. Musique et Sciences Sociales*, special issue 'Sound, Music, Violence', ed. Luis Velasco-Pufleau (2020). Accessed 14 May 2020, https://doi.org/10.4000/transposition.4770.

Hainge, Greg. *Noise Matters: Towards an Ontology of Noise*. New York: Bloomsbury, 2013.

Hajkowski, Thomas. *The BBC and National Identity in Britain, 1922–53*. Manchester: Manchester University Press, 2010.

Hall, Peter. *Cities of Tomorrow*. Malden, MA: Blackwell, 2002.

Hallett, Christine. *Containing Trauma: Nursing Work in the First World War*. Manchester: Manchester University Press, 2009.

Hardy, Dennis and Colin Ward, *Arcadia for All: The Legacy of a Makeshift Landscape*. London: Mansell, 1984.

Harker, Ben. '"The Manchester Rambler": Ewan MacColl and the 1932 Mass Trespass', *History Workshop Journal*, 59 (2005), 219–28.

Harris, Alexandra. *Romantic Moderns: English Writers, Artists and the Imagination from Virginia Woolf to John Piper*. London: Thames and Hudson, 2015.

Harrison, Mark. *The Medical War: British Military Medicine in the First World War*. Oxford: Oxford University Press, 2010.

Haslam, Sara, ed. *Ford Madox Ford: Parade's End Volume III: A Man Could Stand Up*. Manchester: Carcanet, 2011.

Hassan, John. *The Seaside, Health and the Environment in England and Wales Since 1800*. Aldershot: Ashgate, 2003.

Hayes, Nick and Jeff Hill. *Millions Like Us? British Culture in the Second World War*. Liverpool: Liverpool University Press, 1999.

Haythornthwaite, Philip J. *The World War One Source Book*. London: Arms & Amour Press, 1994.

Hendy, David. *Noise: A Human History of Sound and Listening*. New York: Harper Collins, 2013.

Hendy, David. *Public Service Broadcasting*. Basingstoke: Palgrave Macmillan, 2013.

Hendy, David. 'The Great War and British Broadcasting. Emotional Life in the Creation of the BBC', *New Formations*, 82 (2014), 82–99.

Hickman, Clare. 'Cheerfulness and Tranquillity: Gardens in the Victorian Asylum', *Lancet*, 1 (2014), 506–7.

Hickman, Clare. *Therapeutic Landscapes: A History of English Hospital Gardens Since 1800*. Manchester: Manchester University Press, 2013.

Hill, Howard. *Freedom to Roam: The Struggle for Access to Britain's Moors and Mountains*. Ashbourne, Derbyshire: Moorland, 1980.

Hill, Jonathan. *Radio! Radio!* Bampton: Sunrise, 1986.

Hoggart, Richard. *The Uses of Literacy: Aspects of Working-Class Life*. London: Penguin, 2009 (1957).

Holden, Wendy. *Shell Shock: The Psychological Impact of War*. London: Channel 4 Books, 1998.

Holman, Brett. *The Next War in the Air: Britain's Fear of the Bomber, 1908–1941*. Farnham and Burlington: Ashgate, 2014.

Holt, Ann. 'Hikers and Ramblers: Surviving a Thirties' Fashion', *The International Journal of the History of Sport*, 4 (1987), 56–67.

Howes, David, ed. *The Cultural History of the Senses in the Modern Age*. London: Bloomsbury, 2019.

Howkins, Alun. *The Death of Rural England: A Social History of the Countryside Since 1900*. London: Routledge. 2003.

Howorth, Peter. 'The Treatment of Shell Shock: Cognitive Therapy Before Its Time', *Psychiatric Bulletin*, 24 (2000), 225–7.

Humphries, Mark O. and Kellen Kurchinski. 'Rest, Relax and Get Well: A Re-Conceptualisation of Great War Shell Shock Treatment', *War & Society*, 27 (2008), 89–110.

Hunt, Tristram. *Making Our Mark*. London: Campaign to Protect Rural England, 2006.

Hutton, Ronald. *Pagan Britain*. New Haven, CT: Yale University Press, 2013.

Hynes, Samuel. *A War Imagined: The First World War and English Culture*. New York: Atheneum, 1991.

Hynes, Samuel. *The Edwardian Turn of Mind*. Princeton, NJ: Princeton University Press.

Hynes, Samuel. *The Soldiers' Tale: Bearing Witness to Modern War*. New York: Penguin, 1998.

Ingold, Tim. *The Perception of the Environment: Essays on Livelihood, Dwelling, and Skill*. London: Routledge, 2000.

Jackson, Anthony. 'Sound and Ritual', *Man*, 3 (1968), 293–9.

Jackson, Mark. *The Age of Stress: Science and the Search for Stability*. Oxford: Oxford University Press, 2013.

James, David and Philip Tew. *New Versions of Pastoral: Post-Romantic, Modern, and Contemporary Responses to the Tradition*. Madison: Fairleigh Dickinson University Press, 2009.

Jennings, Michael W., Howard Eiland, and Gary Smith, eds. *Walter Benjamin: Selected Writings*, vol. 2: Part 2, *1931–1934*. Cambridge, MA: Harvard University Press, 2005.

Johnson, Paul, ed. *Twentieth Century Britain: Economic, Social and Cultural Change*. Harlow: Longman, 1994.

Jolly, W. P. *Oliver Lodge*. Rutherford: Fairleigh Dickinson University Press, 1975.

Jones, Derek. *Microphones and Muddy Boots: A Journey into Natural History Broadcasting*. Newton Abbot: David & Charles, 1987.

Jones, Edgar. '"An Atmosphere of Cure": Frederick Mott, Shell Shock and the Maudsley', *History of Psychiatry*, 25 (2014), 412–21.

Jones, Edgar. 'Shell Shock at Magull and the Maudsley: Models of Psychological Medicine in the UK', *Journal of the History of Medicine and Allied Sciences*, 65 (2010), 368–95.

Jones, Edgar. 'War Neuroses and Arthur Hurst: A Pioneering Medical Film about the Treatment of Psychiatric Battle Causalities', *Journal of the History of Medicine and Allied Sciences*, 67 (2012), 366–9.

Jones, Edgar and Simon Wessely. 'Psychiatric Battle Casualties: An Intra- and Interwar Comparison', *British Journal of Psychiatry*, 178 (2001), 242–7.

Jones, Edgar and Simon Wessely. *Shell Shock to PTSD: Military Psychiatry from 1900 to the Gulf War*. Hove: Psychology Press, 2005.

Jones, Edgar, Nicola Fear, and Simon Wessely. 'Shell Shock and Mild Traumatic Brain Injury: A Historical Review', *American Journal of Psychiatry*, 164 (2007), 1641–5.

Jones, Edgar, Robin Woolven, Bill Durodié, and Simon Wessely. 'Civilian Morale During the Second World War: Responses to Air Raids Re-Examined', *Social History of Medicine*, 17 (2004), 463–79.

Kahn, Douglas. *Noise, Water, Meat: A History of Sound in the Arts*. Cambridge, MA: MIT Press, 2001.

Keegan, John. *The Face of Battle*. New York: Penguin, 1978.

Kelly, Brendan. *'He Lost Himself Completely': Shell Shock and Its Treatment at Dublin's Richmond War Hospital, 1916–1919*. Dublin: Liffey Press, 2014.

Kendall, Judy. *Edward Thomas, Birdsong and Flight*. London: Cecil Woolf, 2014.

Kenney, William Howland. *Recorded Music in American Life: The Phonograph and Popular Memory, 1890–1945*. New York: Oxford University Press, 2003.

Kern, Stephen. *The Culture of Time and Space 1880–1918*. Cambridge, MA: Harvard University Press, 2003.

Kimmel, Michael S. and Amy Aronson, eds. *Men and Masculinities*, vol. 1. Santa Barbara: Clio, 2004.

Lacey, Kate. *Listening Publics: The Politics and Experience of Listening in the Media Age*. Cambridge: Polity Press, 2013.

Langhamer, Claire. *Women's Leisure in England, 1920–60*. Manchester: Manchester University Press, 2000.

Leed, Eric J. *No Man's Land: Combat and Identity in World War I*. Cambridge: Cambridge University Press, 1979.

Leese, Peter. *Shell Shock: Traumatic Neurosis and the British Soldiers of the First World War*. New York: Palgrave Macmillan, 2002.

LeMahieu, D. L. *A Culture for Democracy: Mass Communication and the Cultivated Mind in Britain Between the Wars*. Oxford: Clarendon, 1988.

Lerner, Paul. 'Psychiatry and Casualties of War in Germany, 1914–18', *Journal Contemporary History*, 35 (2000), 13–28.

Lerner, Paul. *Hysterical Men: War, Psychiatry, and the Politics of Trauma in Germany, 1890–1930*. Ithaca, NY: Cornell University Press, 2008.

Lévi-Strauss, Claude. *Totemism*. London: Merlin Press, 1964.

Lewis-Stempel, John. *Where the Poppies Blow: The British Soldier, Nature, the Great War*. London: Weidenfeld & Nicolson, 2016.

Lloyd, A. L. *Folk Song in England*. St. Albans: Paladin, 1975.

Logie Baird, Iain. 'Capturing the Song of the Nightingale', *Science Museum Journal*, 4 (2015). Accessed 6 September 2017, http://journal.sciencemuseum.ac.uk/browse/issue-04/capturing-the-song-of-the-nightingale/.

Loughran, Tracey. 'Hysteria and Neurasthenia in Pre-War Medical Discourse and in Histories of Shell-Shock', *History of Psychiatry*, 19 (2008), 25–46.

Mabey, Richard. *Flora Britannica*. London: Chatto & Windus, 1997.

Mabey, Richard. *Whistling in the Dark: In Pursuit of the Nightingale.* London: Sinclair-Stevenson, 1993.

MacCulloch, Diarmaid. *Silence: A Christian History.* London: Allen Lane, 2013.

Macdonald, Helen. "'What Makes You a Scientist Is that Way You Look at Things": Ornithology and the Observer 1930–1955', *Studies in History and Philosophy of Biological and Biomedical Sciences*, 33 (2002), 53–77.

Mandler, Peter. 'Against "Englishness": English Culture and the Limits to Rural Nostalgia, 1850–1940', *Transactions of the Royal Historical Society*, 7 (1997), 155–75.

Mansell, James G. 'Musical Modernity and Contested Commemoration at the Festival of Remembrance 1923–1927', *Historical Journal*, 52 (2009), 433–54.

Mansell, James G. *The Age of Noise in Britain: Hearing Modernity.* Urbana: University of Illinois Press, 2017.

Marsh, Jan. *Back to the Land: The Pastoral Impulse in England from 1880–1914.* London: Quartet, 1982.

Marwick, Arthur. 'Middle Opinion in the Thirties: Planning, Progress and Political "Agreement"', *English Historical Review*, 79 (1964), 285–98.

Marwick, Arthur. *The Deluge: British Society and the First World War.* London: Palgrave Macmillan, 2006.

Marx, Leo. *The Machine in the Garden: Technology and the Pastoral Ideal in America.* New York: Oxford University Press, 1967.

Matless, David. *Landscape and Englishness.* London: Reaktion, 1998.

Matless, David. 'Nature, the Modern and the Mystic: Tales from Early Twentieth Century Geography', *Transactions of the Institute of British Geographers*, 16 (1991), 272–86.

Matthews, David. 'The Music of English Pastoral', in *Town and Country*, ed. A. Barnett and R. Scruton (London: Jonathan Cape, 1998), 81–90.

McIntyre, Ian. *The Expense of Glory: Life of John Reith.* London: Harper Collins, 1993.

McIvor, Arthur. 'Employers, the Government, and Industrial Fatigue in Britain, 1890–1918', *British Journal of Industrial Medicine*, 44 (1987), 724–32.

McIvor, Arthur. 'Manual Work, Technology and Industrial Health, 1918–39', *Medical History*, 31 (1987), 160–89.

Merleau-Ponty, Maurice. *Phenomenology of Perception*, trans. C. Smith. London: Routledge, 1992.

Merriman, Peter. ' "Respect the Life of the Countryside": The Country Code, Government and the Conduct of Visitors to the Countryside in Post-War England and Wales', *Transactions of the Institute of British Geographers*, 30 (2005), 336–50.

Miller, Simon. 'Urban Dreams and Rural Reality: Land and Landscape in English Culture, 1920–45', *Rural History*, 6 (1995), 89–102.

Moore, Lucinda. 'Music and Morale: Lena Ashwell and the Healing Power of Concerts at the Front', 18 July 2014. Accessed 6 April 2017, http://blog.maryevans.com/2014/07/music-morale-lena-ashwell-and-the-healing-power-of-concerts-at-the-front.html.

Moore-Colyer, R. J. 'Great Wen to Toad Hall: Aspects of the Urban-Rural Divide in Inter-War Britain', *Rural History*, 10 (1999), 105–24.

Morat, Daniel. *Sounds of Modern History: Auditory Cultures in 19th and 20th Century Europe.* New York: Berghahn, 2014.

Morus, Iwan Rhys, ed. *Bodies/Machines.* Oxford: Berg, 2002.

Moss, Stephen. *A Bird in the Bush: A Social History of Birdwatching.* London: Aurum, 2004.

Mosse, George L. *Fallen Soldiers: Reshaping the Memory of the World Wars.* New York: Oxford University Press, 1991.

Mugglestone, Lynda. 'Rethinking the Birth of an Expression: Keeping Calm and "Carrying On" in World War One', *English Words in Wartime*, 2 August 2016. Accessed 2 September 2016, https://wordsinwartime.wordpress.com/2016/08/02/rethinking-the-birth-of-an-expression-keeping-calm-and-carrying-on-in-world-war-one/.

Mugglestone, Lynda. ' "That Siren Call . . ." The Diverse Language of Air-Raid Precautions in 1916', *English Words in Wartime*, 17 May 2016. Accessed 5 September 2017, https://wordsinwartime.wordpress.com/2016/05/17/that-siren-call-the-diverse-language-of-air-raid-precuations-in-1916/.

Mynott, Jeremy. *Birdscapes: Birds in Our Imagination and Experience*. Princeton, NJ: Princeton University Press, 2009.

Nicholas, Siân. *The Echo of War: Home Front Propaganda and the Wartime BBC, 1939–45.* Manchester: Manchester University Press, 1996.

Niemann, Derek. *Birds in a Cage: Warburg, Germany, 1941. Four P.O.W. Birdwatchers, the Unlikely Beginnings of British Wildlife Conservation.* London: Short Books, 2013.

Nott, James. *Going to the Palais: A Social and Cultural History of Dancing and Dance Halls in Britain, 1918–1960.* Oxford: Oxford University Press, 2015.

Novak, David and Matt Sakakeeny, eds. *Keywords in Sound.* Durham, NC: Duke University Press, 2015.

O'Shea, Helen. 'Defining the Nation and Confining the Musician: The Case of Irish Traditional Music', *Music and Politics*, 3/2 (2009), https://doi.org/10.3998/mp.9460 447.0003.205.

Overy, Richard. *The Bombing War: Europe 1939–1945.* London: Penguin, 2014.

Overy, Richard. *The Morbid Age: Britain Between the Wars.* London: Allen Lane, 2009.

Palmer, Roy. *The Sound of History: Songs and Social Comment.* Oxford: Oxford University Press, 1988.

Peacock, Charlotte. *Into the Mountain: A Life of Nan Shepherd.* Cambridge: Galileo, 2018.

Pegler, Martin. *Soldiers' Songs and Slang of the Great War.* London: Osprey, 2014.

Peters, John Durham. *Speaking into the Air: A History of the Idea of Communication.* Chicago: University of Chicago Press, 1999.

Picker, John M. *Victorian Soundscapes.* Oxford: Oxford University Press, 2003.

Pinch, Trevor and Karin Bijsterveld, eds. *The Oxford Handbook of Sound Studies.* New York: Oxford University Press, 2012.

Prynn, David. 'The Clarion Clubs, Rambling and the Holiday Associations in Britain Since the 1890s', *Journal of Contemporary History*, 11 (1976), 65–77.

Pugh, Martin. *We Danced All Night: A Social History of Britain Between the Wars.* London: Vintage, 2009.

Raymond Williams, *Problems in Materialism and Culture.* London: Verso, 1980.

Readman, Paul. *Storied Ground: Landscape and the Shaping of English National Identity.* Cambridge: Cambridge University Press, 2018.

Reid, Fiona. *Broken Men: Shell Shock, Treatment and Recovery in Britain: 1914–1930.* London: Continuum, 2010.

Reid, Fiona. *Medicine in First World War Europe: Soldiers, Medics, Pacifists.* London: Bloomsbury, 2017.

Revill, George. 'Music and the Politics of Sound: Nationalism, Citizenship, and Auditory Space', *Environment and Planning D: Society and Space*, 18 (2000), 597–613.

Reynolds, David. *The Long Shadow: The Great War and the Twentieth Century.* London: Simon and Schuster, 2014.

Reznick, Jeffrey S. *Healing the Nation: Soldiers and the Culture of Caregiving in Britain During the Great War*. Manchester: Manchester University Press, 2005.

Rice, Tom. *Hearing and the Hospital: Sound, Listening, Knowledge and Experience*. Hereford: Sean Kingston, 2013.

Richards, Jeffrey. *Films and British National Identity: From Dickens to Dad's Army*. Manchester: Manchester University Press, 1997.

Ritschel, Daniel. *The Politics of Planning: The Debate on Economic Planning in the Britain in the 1930s*. Oxford: Clarendon Press, 1997.

Ritvo, Harriet. *The Animal Estate: The English and Other Creatures in the Victorian Age*. Cambridge, MA: Harvard University Press, 1989.

Roper, Michael. *The Secret Battle: Emotional Survival in the Great War*. Manchester: Manchester University Press, 2009.

Rose, Nikolas. *The Psychological Complex: Psychology, Politics and Society in England 1869–1939*. London: Routledge & Kegan Paul, 1985.

Rose, Sonya O. *Which People's War? National Identity and Citizenship in Wartime Britain 1939–1945*. New York: Oxford University Press, 2004.

Ross, Alex. *The Rest Is Noise: Listening to the Twentieth Century*. London: Harper Perennial, 2009.

Rothenberg, Mark. *Why Birds Sing*. London: Penguin, 2006.

Samuel, Raphael, ed. *Patriotism: The Making and Unmaking of British National Identity III: National Fictions*. London: Routledge, 1989.

Samuel, Raphael. *Theatres of Memory*, vol. 1, *Past and Present in Contemporary Culture*. London: Verso, 1994.

Samuel, Raphael. *Theatres of Memory*, vol. 2, *Island Stories: Unravelling Britain*. London: Verso, 1998.

Saunders, Nicholas J. and Paul Cornish, eds. *Modern Conflict and the Senses*. Abingdon, Routledge, 2017.

Saylor, Eric. '"It's not Lambkins Frisking at All"': English Pastoral Music and the Great War', *The Musical Quarterly*, 91 (2009), 39–59.

Saylor, Eric. *English Pastoral Music: From Arcadia to Utopia, 1900–1955*. Urbana: University of Illinois, 2017.

Scannell, Paddy. *Radio, Television and Modern Life*. Oxford: Blackwell, 1996.

Scannell, Paddy and David Cardiff. *A Social History of British Broadcasting: 1922–1939, Serving the Nation*. Oxford: Basil Blackwell, 1991.

Schafer, R. Murray. *Soundscape: Our Sonic Environment and the Tuning of the World*. Rochester: Destiny Books, 1994.

Schmidt, Leigh Eric. *Hearing Things: Religion, Illusion, and the American Enlightenment*. Cambridge, MA: Harvard University Press, 2000.

Schwartz, Hillel. *Making Noise: From Babel to the Big Bang & Beyond*. Brooklyn: Zone, 2011.

Searle, Geoffrey. *A New England? Peace and War, 1886–1918*. Oxford: Oxford University Press, 2004.

Serres, Michel. *The Five Senses: A Philosophy of Mingled Bodies*, trans. M. Sankey and P. Cowley. London: Bloomsbury, 2008.

Sheail, John. *Rural Conservation in Interwar Britain*. Oxford: Oxford University Press, 1981.

Shephard, Ben. *A War of Nerves: Soldiers and Psychiatrists, 1914–1994*. London: Jonathan Cape, 2000.

Shoard, Marion. *A Right to Roam*. Oxford: Oxford University Press, 1999.

Showalter, Elaine. *The Female Malady: Women, Madness and English Culture 1830–1980*. London: Virago, 1987.

Sissons, Dave, Terry Howard, and Roly Smith. *Clarion Call: Sheffield's Access Pioneers*. Sheffield: Clarion Call, 2017.

Smith, Bruce R. *The Acoustic World of Early Modern England*. Chicago: University of Chicago, 1999.

Smith, Mark M. *How Race Is Made: Slavery, Segregation, and the Senses*. Chapel Hill: University of North Carolina Press, 2006.

Smith, Mark M. *The Smell of Battle, the Taste of Siege: A Sensory History of the Civil War*. New York: Oxford University Press, 2015.

Smith, Mark M. 'Sound—So What?', *The Public Historian*, 37 (2015), 132–44.

Snaith, Anna. *Sound and Literature*. Cambridge: Cambridge University, 2020.

Solnit, Rebecca. *Wanderlust: A History of Walking*. London: Granta, 2014.

Sommer, Marianne. 'Animal Sounds Against the Noise of Modernity and War: Julian Huxley (1887–1975) and the Preservation of the Sonic World Heritage', *Journal of Sonic Studies*. Accessed 9 September 2017, https://www.researchcatalogue.net/view/325229/325230.

Stephenson, Tom. *Forbidden Land*. Manchester: Manchester University Press, 1989.

Sterne, Jonathan. *The Audible Past: The Cultural Origins of Sound Reproduction*. Durham, NC: Duke University Press, 2003.

Stiles, Anne. 'The Rest Cure, 1873–1925', *Britain, Representation, and Nineteenth-Century History*. Accessed 1 August 2017, http://www.branchcollective.org/?ps_articles=anne-stiles-the-rest-cure-1873-1925.

Stoever, Jennifer Lynn. *The Sonic Color Line: Race and the Cultural Politics of Listening*. New York: New York University Press, 2016.

Street, Seán. *The Memory of Sound: Preserving the Sonic Past*. New York, Routledge, 2015.

Street, Seán. *The Poetry of Radio: The Colour of Sound*. Abingdon: Routledge, 2012.

Taylor, A. J. P. *English History 1914–45*. Oxford: Oxford University Press, 1965.

Tebbutt, Melanie. 'Landscapes of Loss: Moorlands, Manliness and the First World War', *Landscapes*, 5 (2004), 114–28.

Tebbutt, Melanie. 'Rambling and Manly Identity in Derbyshire's Dark Peak, 1880s–1920s', *The Historical Journal*, 49 (2006), 1125–53.

Terraine, John. *White Heat: The New Warfare 1914–18*. London: Sidgwick and Jackson, 1982.

Thomas, Keith. *Man and the Natural World*. Oxford: Oxford University Press, 1996.

Thompson, E. P. *The Making of the English Working Class*. London: Victor Gollancz, 1963.

Thompson, E. P. 'History From Below', *Times Literary Supplement*, 7 April 1966, 279–80.

Thompson, Emily. *The Soundscape of Modernity: Architectural Acoustics and the Culture of Listening in America, 1900–1933*. Cambridge, MA: MIT Press, 2004.

Thompson, John B. *The Media and Modernity: A Social Theory of the Media*. Cambridge: Polity Press, 2001.

Thompson, Marie. *Beyond Unwanted Sound: Noise, Affect and Aesthetic Moralism*. London: Bloomsbury, 2017.

Thomson, Mathew. *Psychological Subjects: Identity, Culture, and Health in Twentieth-Century Britain*. Oxford: Oxford University Press, 2006.

Tullett, William. *Smell in Eighteenth-Century England: A Social Sense*. Oxford: Oxford University Press, 2019.

Urry, Jon. 'The Tourist Gaze "Revisited," ' *American Behavioural Scientist*, 36 (1992), 172–86.

Van der Kloot, William. 'Lawrence Bragg's Role in the Development of Sound-Ranging in World War I', *Notes and Records of The Royal Society*, 59 (2005), 273–84.

Wainwright, Martin, ed. *Gleaming Landscape. 100 Years of the Guardian Country Diary*. London: Arum, 2006.

Walker, Helen. 'The Popularisation of the Outdoor Movement, 1900–1940', *The British Journal of Sports History*, 2 (1985), 140–53.

Watson, Alexander. *Enduring the Great War: Combat, Morale and Collapse in the German and British Armies, 1914–1918*. Cambridge: Cambridge University Press, 2008.

Webb, Thomas. ' "Dottyville"—Craiglockhart War Hospital and Shell-Shock Treatment in the First World War', *Journal of the Royal Society of Medicine*, 99 (2006), 342–6.

Welch, David. *Germany, Propaganda and Total War, 1914–1918*. New Brunswick: Rutgers University Press, 2000.

White, Jerry. *Zeppelin Nights: London in the First World War*. London: Vintage, 2015.

Williams, Raymond. *Keywords: A Vocabulary of Culture and Society*. London: Fontana Paperbacks, 1990.

Williams, Raymond. *Problems in Materialism and Culture*. London: Verso, 1980.

Winter, Denis. *Death's Men: Soldiers of the Great War*. London: Penguin, 1979.

Winter, Jay. 'Shell-Shock and the Cultural History of the Great War', *Journal of Contemporary History*, 35 (2000), 7–11.

Winter, Jay. *The Great War and the British People*. London: Palgrave Macmillan, 2003.

Winter, Jay, ed. *The Cambridge History of the First World War*. Cambridge: Cambridge University Press, 2014.

Winter, Jay, Geoffrey Parker, and Mary Habeck, eds. *The Great War and the Twentieth Century*. New Haven, CT: Yale University Press, 2006.

Zeepvat, Charlotte. *Before Action: William Noel Hodgson and the 9th Devons. A Story of the Great War*. Barnsley: Pen & Sword, 2015.

Zweiniger-Bargielowska, Ina. *Managing the Body: Beauty, Health, and Fitness in Britain, 1880–1939*. Oxford: Oxford University Press, 2010.

Index